Jodenir Calixto Teixeira

Sugarcane expansion in the Ivinhema River Basin - MS (2003-2015)

Jodenir Calixto Teixeira

Sugarcane expansion in the Ivinhema River Basin - MS (2003-2015)

Socio-spatial effects

ScienciaScripts

Imprint

Cover image: www.ingimage.com

This book is a translation from the original published under ISBN 978-3-330-76830-7.

Publisher:
Sciencia Scripts
is a trademark of
Dodo Books Indian Ocean Ltd. and OmniScriptum S.R.L publishing group

120 High Road, East Finchley, London, N2 9ED, United Kingdom
Str. Armeneasca 28/1, office 1, Chisinau MD-2012, Republic of Moldova, Europe
Managing Directors: Ieva Konstantinova, Victoria Ursu
info@omniscriptum.com

Printed at: see last page
ISBN: 978-620-8-59531-9

SUMMARY

PREFACE

It is with great satisfaction that I write the preface to this book by Professor Jodenir Calixto Teixeira, from the Federal University of Mato Grosso do Sul, Três Lagoas Campus.

Analyzing the effects of sugarcane expansion on the Ivinhema river basin, in Mato Grosso do Sul, was the main objective of the doctoral research that forms the basis for this book.

The author set out to demonstrate the socio-spatial effects of the expansion of the sugar-alcohol sector in the Ivinhema river basin, showing that, for the most part, the region's most vulnerable actors were not taken into account. Although the sugar-energy sector is an important economic activity, its social and environmental impacts have been more negative than positive for the majority of the population living in the municipalities that make up the Ivinhema River basin.

In addition to surveying and analyzing bibliography directly and indirectly related to the subject of the research, the author carried out exhaustive fieldwork in rural establishments in the Ivinhema River basin, interviewed businessmen in the sugar-alcohol sector and key people linked to state and federal public bodies operating in the region studied.

The book is richly illustrated with maps of land use and occupation in the Ivinhema river basin, which were drawn up based on the interpretation of images from the *Landsat* Satellite, showing the evolution of sugar cane cultivation in the basin between 2003 and 2015.

In the first two chapters of the book, the author makes consistent analyses and discussions about the process of modernization of Brazilian agriculture, as well as the expansion of sugar cane cultivation and its socio-spatial effects, with an emphasis on the period after 1975, when the Brazilian government launched the National Alcohol Program (PROÀLCOOL).

In the other three chapters that make up the book, the author analyzes sugarcane expansion in the agrarian context of the state of Mato Grosso do Sul, characterizes the Ivinhema river basin, with an emphasis on the effects caused by the expansion of the sugar-alcohol sector in the agrarian space and agriculture of that river basin.

The book provides a consistent analysis of the process of agricultural modernization and the socio-spatial effects generated by the expansion of the sugar-alcohol sector in Brazil, in Mato Grosso do Sul and in the watershed studied.

The author points out that the use of Brazilian agricultural land is not strictly regulated, which means that the most powerful economic groups have a strong influence on the spatial (re)arrangement of agricultural production. The most profitable types of exploitation and those that arouse the greatest interest from strong economic groups are usually also more stimulated by the government, which is the case of the sugar-alcohol sector, which for most of the first decade of the 21st century enjoyed special treatment. Vast resources were made available by the National Bank for Economic and Social Development (BNDES) for the installation of new mills, the modernization and expansion of industrial plants already installed, the implantation of new crops and the replanting of sugar cane and investments in technologies for the mills to implement structures linked to the cogeneration of electricity.

In the 2000s, there was a significant expansion of the sugar-alcohol sector towards the Midwest of the country, especially in Mato Grosso do Sul, where the installation of new mills and the planting of sugarcane crops were concentrated mainly in the upper part of the Ivinhema river basin, where soils with greater natural fertility predominate, as Prof. Jodenir has analyzed very well in this book. Around 80% of the plants installed in Mato Grosso do Sul are located in municipalities that are part of this hydrographic basin.

The empirical study was carried out with care and is of great importance in the work as a whole, both because it clarified several doubts, even leading to the refutation of initially formulated hypotheses, and because it provided security for the preparation of the maps that show the changes that have occurred in the use and occupation of agricultural land in the hydrographic basin studied.

When he began the research that led to this book, Prof. Jodenir formulated the hypothesis that food production areas were being replaced by sugarcane plantations in the area of the Ivinhema River basin - MS. During the course of the research, the author found that sugarcane plantations had expanded mainly in areas previously occupied by pastures and that, in recent years, areas cultivated with sugarcane have begun to give way to the cultivation of grains, mainly soybeans and corn. The production of basic foodstuffs in that area, both in the past and in the present, is carried out by small family producers, who have always faced and continue to face great difficulties in obtaining an income that guarantees their livelihood.

By studying the effects of the expansion of the sugar-energy sector in the Ivinhema River Basin, the author manages to establish the connections between this phenomenon and broader processes occurring on the scales of Mato Grosso do Sul and the country, making use of consultations and analyses of relevant works, as well as consistent field research carried out in the Ivinhema River Valley.

Reading the book, besides being pleasant, thought-provoking and enlightening, allows the reader to broaden their knowledge of agricultural dynamics, the modernization of agriculture and the expansion of the sugar-alcohol sector in Brazil, in Mato Grosso do Sul and in the area studied, at the same time as providing access to relevant data and information on the Ivinhema River Basin, as well as on the processes and phenomena that have resulted in changes in the landscape of this important area of Mato Grosso do Sul over time, including through the rich cartographic material included in the book.

Antonio Nivaldo Hespanhol

Presidente Prudente (SP), March 2017.

DEDICATORY

To my parents, Margarida Rufino Calixto and Jesus dos Santos Teixeira, who, in the simplicity in which they always lived, were "warriors" and proved to those who doubted that poor people's children can also study. To my beloved wife, Glaucimar de Queiroz Vasques Teixeira, for her unconditional support throughout our history and to our son Douglas Vasques Teixeira, the treasure God placed in our lives and the main reason for my struggles.

ACKNOWLEDGMENTS

As a Christian, my faith has always accompanied me throughout my journey and made me believe that it would be possible to achieve all my goals, combining my spirituality with my efforts and dedication. That's why I thank God first and foremost, from where all my inner strength comes, who made a boy with his feet on the ground in the past become a doctor today.

And to those who collaborated directly or indirectly in carrying out the research that gave rise to this book, my eternal gratitude goes out in particular:

- To Prof. Dr. Antonio Nivaldo Hespanhol, advisor and friend, who directed the paths of this research, a professional from whom I will take only good examples;

- To the Pro-Rectory of Research and Postgraduate Studies - PROPP/UFMS, for granting me full leave during my doctorate;

- To Prof. Dr. Messias Modesto dos Passos, Prof. Dr. Sedeval Nardoque, Prof. Dr. Marcos Aurélio Saquet and Prof. Dr. Wagner Luiz Lourenzani for their pertinent contributions to the defense panel, which guided this work to its conclusion;

- To the professors of the post-graduate course in Geography at FCT-UNESP, especially those who taught the subjects I took and who provided important reflections on Geographical Science for this research;

- To the students of the Postgraduate course at FCT-UNESP, especially those entering in 2012 and members of the Regional Dynamics and Agriculture Study Group - GEDRA;

- To the rural producers in the Ivinhema river basin area who welcomed me promptly, answered the questions on the form and granted interviews, which were fundamental in making the research a reality;

- To the technicians of the Agrarian Development and Rural Extension Agency - AGRAER, in the municipalities of Anaurilândia, Nova Andradina, Ivinhema, Deodàpolis and Dourados, for the interviews and data provided;

- To the Mato Grosso do Sul State Secretariat for the Environment, Planning, Science and Technology - SEMAC-MS, for the data provided;

- To the employees of the Aurora, Laguna, Santa Helena, Adecoagro Vale do Ivinhema - Angélica plant and CBAA - Sidrolândia plant, who were willing to meet me and give me interviews;

- To the professors at CPNA-UFMS and CPTL-UFMS who encouraged me to do the course, especially Ary Tavares Rezende Filho, Marcelino de Andrade Gonçalves, Flàvia Akemi Ykuta, Denis Richter, Paulo Duarte Paes and Mara Aline dos Santos Ribeiro, fellow students on the Geography course at CPNA-UFMS, for their support and friendship;
- To my former students and friends, Amorim, Dercilio and Wagner, for giving me accommodation when I was in Nova Andradina and helping me with my fieldwork;
- To former students and friends, Paulo D'Alkmin and Edvan, for their help with fieldwork in the municipalities of Nova Andradina and Taquarussu;
- To my friends Marcelina and Jânio, for their support and accommodation when I returned to Presidente Prudente;
- To my parents (Margarida and Jesus), siblings (Jair, Maria Aparecida and Elizabeth), other family members and friends for their encouragement and prayers;
- To Glaucimar (wife) and Douglas (son) for their support and understanding during my absences.

Thank you all very much!

"Sometimes we feel that what we do is just a drop in the ocean. But the sea would be smaller if it lacked a drop". (Mother Teresa of Calcutta)

SUMMARY

This paper presents the results of an investigation carried out in the area of the Ivinhema river basin, in the state of Mato Grosso do Sul, with the aim of analyzing the expansion of sugarcane production in this area over the last twelve years and its effects on the area. In addition to bibliographical research, data was collected from secondary sources, interviews were conducted with key people and forms were applied to rural producers in 24 of the 25 municipalities that make up the Ivinhema river basin. The area of the hydrographic basin studied has seen a strong expansion of sugar-alcohol activity in recent years, especially in the 2000s, when several sugar and alcohol plants were set up and sugarcane plantations began to occupy vast areas. The expansion of sugarcane cultivation took place mainly on areas previously occupied by pastures and, to a lesser extent, on spaces that were occupied by other large crops, mainly soybeans and corn. At the end of the 2000s, the sugar-alcohol sector went into crisis and today there is a reduction in the areas under sugar cane cultivation and a greater expansion of soybean cultivation, although the areas under pasture continue to shrink. It was concluded that, in the last twelve years, there have been significant changes in the agricultural space of the Ivinhema river basin, but the crops that are of interest to agribusiness have remained predominant, with sugar cane, soybeans and corn standing out. Other crops, especially food crops, are grown on much smaller areas, almost exclusively in small rural establishments.

Keywords: Ivinhema river basin. Agricultural development. Agribusiness. Crops. Sugarcane.

INTRODUCTION

In recent decades, sugar cane cultivation in Brazil has been more representative than in its entire history. This crop, which had already been grown in the Northeast since colonial times, expanded to the Center-South at a rapid pace, especially since the 1970s, with the advent of the National Alcohol Program (PROÀLCOOL) and, in the 2000s, moved towards other areas of this region, including Mato Grosso do Sul.

For Castro *et. al.* (2010), we can divide the sugar cane cycle in Brazil into three phases: when it arrived and expanded in the Northeast during the colonization period; when it expanded as a result of PROÀLCOOL in the 1970s in the Center-South, especially in the Southeast, and when it expanded towards the north and northwest of the Center-South, mainly in the 2000s, associated with the global energy crisis, linked to fossil fuels.

In this last phase, production surpassed all previous phases and thousands of hectares were transformed into sugarcane plantations, which, according to Souza (2010), caused a series of socio-environmental impacts in the agrarian space, such as: greater land concentration, territorial disputes in food production areas and accentuated exploitation of the workforce.

The secondary data, organized in this work, points to a reduction in the area of food crops in Mato Grosso do Sul and an increase in the area of sugar cane, which, a *priori,* made us think that these crops were being replaced by sugar cane.

In Brazil, there is no policy regulating the use of agricultural land, so the productive space is configured according to the interests of certain groups, in other words, the power of capital determines the spatial (re)arrangement of production. The productive spatial organization takes place according to the most profitable type of production at a given historical moment, obviously favouring the capitals linked to these productive sectors.

The federal government's attempt to organize sugarcane production, with the creation of the Agroecological Zoning of Sugarcane (ZAE), does not solve the problem of replacing food production with sugarcane, since the areas earmarked for zoning this production are generally large food producers.

On this point, Souza (2010) shows that 85% of the country's sugarcane production is in the Center-South region, a major food producer, with the Southeast region standing out, especially the state of São Paulo, with around 60% of sugarcane production.

In fact, what guides the process of territorialization of sugarcane production is the possibility of obtaining more and more income from this activity, which, according to Souza (2010), can reorder the agrarian space of the regions that are part of the national/international economy as major food producers. The easier it is for the region to expand, the more people will seek to occupy it with this crop, allowing for greater profit margins and income.

In this context, sugar cane cultivation in the Midwest in the 2000s took place at an accelerated pace and, in the case of Mato Grosso do Sul, was concentrated mainly in the Ivinhema river basin area.

As a result, this region, which was integrated into the national economy as a producer of grain and livestock, became the target of the agro-industrial sugarcane agribusiness in the 1970s, due to the growing number of sugarcane mills.

This basin covers 4.64 million hectares, corresponding to 5.27% of the Paranà basin in Brazil. It lies entirely in the state of Mato Grosso do Sul and stands out for having the highest agricultural production in the state, especially in its upper part.

This fact caught our attention and prompted our research, because the advance of sugar cane is much greater in the upper part of the basin, with more fertile soils, where grain production predominates, than in the lower part, with less fertile soils, where cattle ranching is prominent. This led us to hypothesize that there was a substitution of sugarcane plantations, i.e. the areas traditionally planted with grain and food were being taken over by sugarcane plantations, and that this was getting worse with the installation of new mills.

In view of the above, we set out to answer the following questions: what are the effects of the expansion of the sugar-alcohol sector in the Ivinhema river basin? Are all the actors involved in this process being considered? What are the impacts on the economy of these municipalities? Is it, in fact, the replacement of food crops with sugar cane?

The main objective of the research was to analyze the expansion of sugarcane plantations in the Ivinhema river basin over the last twelve years, as well as its effects on this part of the state of Mato Grosso do Sul.

The specific objectives of the research were as follows: a) to demonstrate the changes that have taken place in the agrarian space of the Ivinhema river basin as a result of the expansion of sugarcane plantations; b) to demonstrate the productive evolution of this agrarian space during this period; c) to identify the people involved in the process of expansion of the sugar-alcohol sector in the Ivinhema river basin.

In order to achieve the objectives set, the following methodological procedures were adopted: bibliographical survey on agricultural production in Brazil and Mato Grosso do Sul, with a focus on sugarcane expansion; bibliographical survey on the region studied; collecting data on agricultural production in the municipalities that belong to the Ivinhema river basin, from the Brazilian Institute of Geography and Statistics (IBGE), the National Institute for Space Research (INPE), the Agency for Agrarian Development and Rural Extension (AGRAER), the National Institute for Colonization and Agrarian Reform (INCRA) and the Mato Grosso do Sul State Secretariat for the Environment, Planning, Science and Technology (SEMAC-MS); carrying out field research in part of the rural establishments in the region, with photographic records and application of a form; carrying out interviews with rural producers and in part of the mills installed in the area; organizing the data collected in graphs and tables; mapping the basin area using satellite images, showing land use and occupation in 2003, 2006, 2009, 2013 and 2015; data analysis and final drafting based on the bibliographical references, field research and mapping.

To prepare this work, we proceeded as follows:

a) **Analysis of the bibliography consulted:** after the bibliographical survey, we read and summarized the main works on the subject, especially those focused on discussing the territorial transformations of the agrarian space, with an emphasis on the role of sugarcane expansion in this process. This theoretical background was extremely important for understanding the data collected in the field, discussing it and writing the paper.

b) **Field research:** As an important tool for geographers, empirical research allowed us to make on-the-spot observations about changes in the agricultural space under study and was fundamental to achieving the objectives set.

We surveyed 75 rural establishments, in which we collected data on the production structure and the profile of the producer, by conducting interviews and applying forms containing objective and open-ended questions. This stage of the work was important because it gave us direct contact with the area under study and the observations made allowed us to compare the data collected (primary and secondary), as well as providing us with important photographic records. The establishments surveyed are located in 24 of the 25 municipalities in the Ivinhema basin.

In order to gather information on agricultural production in the basin area and on rural producers, we also conducted interviews with the Agricultural Development and Rural Extension Agency (AGRAER) in the municipalities of Anaurilândia, Nova Andradina, Ivinhema, Deodàpolis and Dourados; the Mato Grosso do Sul State Secretariat for the Environment, Planning, Science and Technology (SEMAC-MS), in Campo Grande and; the National Institute for Colonization and Agrarian Reform (INCRA), in Dourados.

In order to obtain information about the mills installed in the area, we tried to arrange visits to these companies, but we didn't have much success and were only received by five of the 17 in operation until 2014, i.e. 29.4% of the total. Two of them replied to our questionnaire via email and the others did not offer to meet us.

c) **Analysis of primary data:** the data collected through interviews and the application of forms were tabulated, analyzed and contributed to the understanding of important issues throughout the work. The forms used and the interviews helped us form an important database for the analysis.

d) **Secondary data analysis:** In this stage, we proceeded to analyze the data collected, organized in tables and graphs. This data, together with the primary data, forms the quantitative basis of the research which, when systematized, has helped us to analyse the productive dynamics of the region under study, thus fostering theoretical discussion.

e) **Land use and occupation mapping: Land** use and occupation maps of the Ivinhema river basin area were drawn up for the years 2003, 2006, 2009, 2013 and 2015. It was not possible to maintain the three-year interval between them due to the high concentration of clouds in 2012, making it difficult to see the images. To create these maps, we interpreted the satellite images Landsat 5/TM, Landsat 7/ETM, bands 3, 4 and 5 and Landsat 8/OLI, bands 4, 5 and 6 available on the INPE *website* and from the United States Geological Survey. The images were imported and georeferenced using control points that could be identified on the image, and a panchromatic orthorectified image of the area was used as the basis for registering the new satellite images, worked on in the ArcGis GIS® 10. After georeferencing, contrast was applied in the Spring GIS® 5.2.3 to improve image quality, using the histogram equalize option. Based on this contrast, we saved the synthetic image of the satellite images and carried out segmentation and unsupervised classification by region, defining seven classes of land use and occupation. To finalize the maps, we again used ArcGis software® 10.

Therefore, this work originated from the results of the thesis entitled "The socio-spatial effects of sugarcane expansion in the Ivinhema river basin in the state of Mato Grosso do Sul", defended in December 2015, at the Geography Postgraduate Program of the Faculty of Science and Technology (FCT) of the São Paulo State University (UNESP), in Presidente Prudente-SP.

We have structured this work in five chapters, as follows:

In the first, entitled "Modernization of agriculture and rural development in Brazil", we analyse the post-World War II period, when there was greater investment in Brazilian agriculture, culminating in its modernization process from the 1960s onwards. In the text, we demonstrate the main changes that took place in Brazilian agriculture during this period, the priority given to large producers by the government and, consequently, in recent times, the support given to the advance of agribusiness. In this chapter, we also look at the expansion of sugar cane in this context, especially in the 1970s, with the establishment of PROÀLCOOL and the growing need for ethanol production.

This chapter is based on the following authors: Guimarâes (1979), Kageama (1987), Gerardi (1980), Graziano da Silva (1981, 1996, 1997 and 2000), Silva (1983), Delgado (1985), Graziano Neto (1985), Sorj (1986), Brum (1988), Borges *et. al.* (1988), Müller (1989), Magalhâes, *et. al.* (1991), Santos (1994), Abramovay (1996), Gonçalves Neto (1997), Marafon (1998), Coelho (2001), Santos and Silveira (2001), Fernandes (2004); Mendonça (2004 and 2008), Arruzzo (2005), Santana (2005), Teixeira (2005), Balsam (2006), Carvalho and Carrijo (2007), Guanziroli (2006), Oliveira (2006, 2010), Veiga Filho and Ramos (2006), Hespanhol (2007, 2008a and 2008b), Castro (2008), Michellon *et.* al. *al.* (2008), Silva (2008), Teubal (2008), Andrade *et.al.* (2009), Mota and Pessôa (2009), Cleps Jûnior (2010), Monteiro (2010), Souza (2014) and Tonin and Tonin (2014).

In the second chapter, entitled "The trajectory of sugar cane in Brazil", we summarize the historical aspects of this product in the country, from the colonial period to the present day, highlighting some important points for understanding the trajectory of this crop in Brazilian territory, with greater emphasis on the 20th century. In this chapter, we also point out some of the socio-spatial effects of the expansion of sugar cane in Brazil and current trends.

The following works were important for this chapter: Furtado (1961), Fauconnier and Bassereau (1970), Pina (1972), Szmrecsànyi (1979), Silva (1983), Prado Jûnior (1987), Andrade (1988), Magalhâes *et.* al. *al.* (1991), Bray *et. al.* (2000), Thomaz

Júnior (2002 and 2008), Baccarin (2005), Carvalho and Carrijo (2007), Oliveira (2007, 2008 and 2009), WWF-Brasil (2008), Brito (2008), Camargo *et. al. al.* (2008), Cordeiro (2008), Costa (2008), Hespanhol (2008a), Szmrecsànyi *et. al.* (2008), Andrade *et. al.* (2009), Goes and Marra (2009), Mendonça (2009), Segatti (2009), Xavier and Bombardi (2009), Costa *et. al.* (2010), Souza (2010, 2011 and 2014), Xavier (2010), Carvalho and Santos (2012) and Farias and Silva (2012).

In Chapter 3, entitled "The expansion of sugarcane in the agrarian context of the state of Mato Grosso do Sul", we present a summary of the occupation of the area that is now the state of Mato Grosso do Sul, as well as its creation in the 1970s. However, the main objective of the chapter was to demonstrate the evolution of agricultural production in the state and to situate the advance of sugarcane in this context.

For this chapter, we used the following works: Rosa (1962), Rodrigues (1978), Duarte (1989), Teixeira (1989), Campestrini and Guimaraes (1991), Oliveira (1993 and 2009), Arbex and Olic (1996), Guimaraes and Leme (1997), Santos and Silveira (2001), Teixeira (2001), Rezende (2002), Monteiro (2006), Teixeira and Hespanhol (2006), Pereira *et. al. al.* (2007), Santos (2007), Backes (2008), Faria and Frata (2008), Nassar *et. al.* (2008), Carrijo and Miziara (2009), Asevedo and Ribeiro (2010), Azevedo (2010), Castro *et. al.* (2010), Monteiro (2010), Kudlavisk (2011), Santos and Schlindwein (2011) and Menezes (2012).

In Chapter 4, entitled "Characterization of the Ivinhema River Basin in the State of Mato Grosso do Sul", we analyze the importance of research on river basins in the field of Geographical Science and then briefly characterize the area of the Ivinhema River Basin in terms of its geographical aspects. In this chapter we also present the profile of the rural producers and the characteristics of the rural establishments surveyed, based on the fieldwork.

For the preparation of this chapter, in addition to the results of the fieldwork, we used some works that helped us to characterize and understand the space studied, including: Santos (1982), Oliveira, Urchei and Fietz (2000), Castro (2003), Lima (2005), Faria and Frata (2008), Ferreira and Brandao (2011) and Moraes (2011).

In Chapter 5, entitled "Sugarcane expansion in the context of agricultural production in the Ivinhema River basin - MS", we analyze sugarcane expansion in the Ivinhema River basin, based on the systematization of primary and secondary data, as well as our field observations. The secondary data was mainly sourced from the Brazilian Institute of Geography and Statistics (IBGE), the National Institute for Space Research

(INPE), the Mato Grosso do Sul Agricultural Development and Rural Extension Agency (AGRAER) and the Mato Grosso do Sul State Secretariat for the Environment, Planning, Science and Technology of Mato Grosso do Sul (SEMAC-MS), but information obtained from the Union of Bioenergy Producers (UDOP), the Mato Grosso do Sul Bioenergy Association (BIOSUL), the Union of Sugarcane Industries (UNICA) and the Mato Grosso do Sul Federation of Agriculture and Livestock (FAMASUL) was also important. The primary data was collected from the 75 rural establishments surveyed and from the interviews conducted.

In addition, the mapping of land use and occupation in the basin area has helped us to understand the evolution of sugarcane production over the last twelve years.

We also used the following references to prepare this chapter: Faria and Frata (2008), Mizusaki (2009), Monteiro (2010, 2012 and 2014), D'Alkmin Filho and Teixeira (2012), Batista (2014) and Teixeira and Hespanhol (2014).

CHAPTER I: MODERNIZATION OF AGRICULTURE AND RURAL DEVELOPMENT IN BRAZIL

In this chapter, we try to characterize the period of modernization of agriculture in Brazil, especially since the 1970s, when the National Alcohol Program (PROÀLCOOL) was implemented and, consequently, there was a major expansion of sugarcane plantations in the country. We emphasize how the agents of capital acted in the countryside and what their repercussions were on the Brazilian economic scenario, placing the sugar-alcohol sector in this context.

1.1. Modernization of agriculture and its effects on the Brazilian agrarian scene in the post-World War II period

The modernization of agriculture in Brazil, which began in the middle of the 20th century, aimed to increase agricultural production, mainly for export, through the import of machinery, pesticides and chemical fertilizers linked to the so-called "technological package" of the so-called "Green Revolution", promoted mainly by companies and state bodies.

In the post-World War II period, developed countries, especially the United States, were interested in changing the technical basis of agriculture in underdeveloped countries, given the need for greater production of raw materials to be processed by multinational companies.

The "Green Revolution" was also expected to eliminate or at least reduce hunger in the world, as there would be a large production of foodstuffs. However, even with the significant increase in food production, the problems related to hunger persisted (ABRAMOVAY, 1996).

Thus, since the 1960s, Brazil's agrarian space has undergone changes as a result of agricultural modernization, although not with the same intensity or in a homogeneous way across the different parts of the country. This phenomenon has been more concentrated in the South and Southeast, with large landowners (mainly export producers) being privileged and small producers being left on the margins of this policy.

In this sense, Mota and Pessôa (2009, p.1) state:

> These changes are linked to the project to modernize the territory, more specifically to the modernization of the productive structure of the countryside, which has become an unequal process of capital expansion between producers and regions.

The modernization of agriculture required greater investment in infrastructure, mainly to facilitate the movement of goods, people and information (MENDONÇA, 2004).

The use of modern technologies and large-scale production, through the greater integration of capital and the worldwide circulation of production, have become elements of modern agriculture (MOTA and PESSÔA, 2009).

Some data shows the advance in the use of machinery in the Brazilian countryside. If we compare the number of tractors used in rural establishments between 1950 and the mid-2000s in Brazil, according to data from the Brazilian Institute of Geography and Statistics (IBGE), we can see a considerable increase in their number.

In 1950, 8,372 tractors were used. This number rose to 61,338 in 1960; 165,870 in 1970; 323,113 in 1975; 545,205 in 1980; 665,280 in 1985; 803,742 in 1995/96 and fell to 788,053 in 2006, according to data from the Agricultural and Farming Censuses for those years. Over time, there has been a significant expansion in the power of the machines, with tractors with more than 100 horsepower predominating.

According to Graziano Neto (1985), the use of artificial fertilizers and pesticides also increased from the 1960s onwards. According to this author, between 1965 and 1975, there was an increase in the consumption of fertilizers at an average annual rate of 60% and pesticides at an average annual rate of 25%.

The new way of producing meant that the countryside became increasingly independent of natural factors and more dependent on new agricultural inputs. Industry received raw materials from the countryside and supplied the industrialized products (machinery, implements, chemical fertilizers and biocides) needed for large-scale production, in other words, the dependence of agriculture on industry increased (DELGADO, 1985). However, this process has also led to a shift in agricultural production, with the countryside now producing to meet the demand for raw materials from the agro-industrial sector.

From this perspective, agricultural modernization represented the expansion of capitalism in the countryside, restructured to meet the demands of international and industrial capital, guaranteeing crops that generated surpluses in the balance of trade, i.e. the government opted for the model adopted in developed countries, such as the United States, for example. With this, it was believed that the backwardness of the Brazilian countryside could be overcome with the introduction of new production techniques and new equipment, changing working relationships and investing in products that would generate more income for the country.

The advance of modern agriculture altered the country's agricultural frontier, expanding in a central-northern direction, made possible by the subsidized credit made available by the Brazilian state. For Arruzzo (2005), modernization did not take place horizontally across Brazil, as some regions were excluded from the process and others benefited, with the Midwest standing out as an area of intense modernization, with some export products standing out, such as soya.

In this context, small rural producers were left on the sidelines and, according to Graziano Neto (1985), the countryside was divided into "rich people's agriculture", which was modern, and "poor people's agriculture", which was backward.

The model adopted for modernization excluded small producers, as they did not have the resources to buy the inputs needed for the proposed advances, and so some continued to produce only for family consumption and others sold their land and migrated to the cities. This was because small landowners saw no other alternatives, because they were unable to invest in production to keep up with modernization and because of the lack of jobs in the countryside, since modern agriculture required less labour.

The government has left the small producer on the margins of the modernization process and no compensatory public policies have been instituted to assist this section of the population.

As a result, there was an intense displacement of the rural population to the cities, which Santos (1994) calls "forced migration", as these people were "expelled" from the countryside and, between the 1960s and 1980s, Brazil experienced the largest rural-city migration in its history.

The use of mechanization led to permanent workers being laid off because they were no longer needed all year round and their dismissal also relieved the owner of social security payments (BALSAM, 2006, p. 137).

For Mota and Pessôa (2009), this was a terrible process of forced deterritorialization by capital, and this population was part of the social movements fighting for land, generating major conflicts in the countryside in the following decades.

Oliveira (2006) points out that the new pattern of economic development has excluded rural workers from generating employment, reducing their income and, consequently, their standard of living. In this sense, Balsam (2006) reiterates that, from a global perspective and through historical processes, land ownership has been subordinated to capital.

Even so, the model adopted was seen as a symbol of development and economic growth often appeared to be synonymous with economic modernization. But the definitions of modernization are contradictory.

> The concept of the modernization of agriculture varies between the various authors who deal with the subject, some considering only changes in the technical base and others taking into account the entire production process. In the first case, agricultural production is considered modernized when it makes intensive use of equipment and techniques, such as modern machinery and inputs, which allows for greater yields in the production process. Thus, the modernization of agriculture would be synonymous with the mechanization and technification of farming. In the second case, he considers that the concept of modernization cannot be restricted to the equipment used, but must take into account the whole process of changes that have occurred in the social relations of production (TEIXEIRA, 2005, p. 22).

Graziano da Silva (1981) calls this process "painful modernization", because it favoured certain crops and large producers, to the exclusion of small landowners and workers employed in the countryside. For this author, economic modernization was not accompanied by social modernization.

According to Hespanhol (2007, p. 272-273):

> Development only occurs when economic growth is achieved with respect for natural resources and when there is an improvement in the quality of life of the majority of the population, which is very rare, especially in underdeveloped countries, where the actions of the state in correcting the imperfections generated by the market are timid, if not reinforcing these imperfections even more.

However, although contradictory, the results of this process appeared in the decades after the 1960s, with a leap in the production of *commodities* such as soya, for example. The country became a major exporter of grains and there were reductions in traditional crops, while soya expanded considerably due to international demand for this product. If, on the one hand, the production of some products for export has gained more and more space, other crops (mainly for domestic supply) have lost space and had their areas reduced.

Technological innovations have strengthened the domination of capital over labor, consolidating rural companies, weakening the vision of latifundia and reducing the space for small producers.

Modernization along capitalist lines did not take into account the environmental and social repercussions, as it only aimed to increase productivity. With the use of modern techniques, more was produced in a smaller space and using less time, but there was no room for agrarian reform, reinforcing the concentration of land and social exclusion. In addition, the impact on the environment has increased considerably, mainly due to the use of pesticides.

From this perspective, Hespanhol (2008b, p.372) points out:

> Despite the apparent success of the modernization of agriculture, the environmental liabilities resulting from it are very large. The expansion of monocultures and the indiscriminate use of machinery, implements, chemical fertilizers and biocides have compromised the environmental quality of vast areas in developed and underdeveloped countries.

For Brum (1988), the process of modernizing agriculture in Brazil, driven from the outside in, has not produced the expected results, and he adds that in countries where agrarian reform has been implemented, the results have been more satisfactory.

Production serves as an instrument for transforming space, bringing prosperity and decadence. Thus, during its period of modernization, Brazil had moments of intense production and moments of sharp decline, due to the oscillation of the market for certain types of products. The goal was always to produce more and generate more profits, in other words, favoring those products that were well accepted on the domestic and foreign markets, especially soybeans, corn and sugar cane.

> It is therefore important to associate the modernization of agriculture with the space/time category, as it was not homogeneous in time and space. Because it was a process led by the state, some areas were prioritized for capital investment, becoming territories for and of capital in a short space of time (MOTA and PESSÔA, 2009, p. 9).

For Balsam (2006), modernization has been progressive and punctual and has thus made it possible to structural differences in rural space, especially in production.

Gerardi (1980, p. 23) points out some barriers that hinder the progress of the process of modernization of agriculture in Brazil, such as:

a) the producer's assessment of the degree of risk involved in adopting modern techniques and the uncertainty of a guaranteed return due to the high investment costs;

b) the adoption of modern techniques required high investment and depended on capital reserves or financing; and

c) the lack of information or knowledge on the part of producers.

The fact is that agriculture had to adapt to the new demands of the state and certain economic groups. Thus, according to Oliveira (2003), there was a specialization of production on a national scale, with some crops standing out, such as cotton, tobacco, cocoa, coffee, rice, wheat, grapes, soya and sugar cane.

For Graziano da Silva (2000), the advance of the modernization process has also altered the country's land structure, making it even more concentrated and making it even more difficult for small producers to access land. In this sense, Guimarâes (1979) states that the existing inequalities in the countryside have become greater, both in terms of access to land, the infrastructure of properties and the right to credit.

According to Sorj (1986), the establishment of the military regime in 1964 created favorable conditions for the strengthening of capitalism in the Brazilian countryside, strengthening the expansion of monocultures, centered on large foreign, state and national capital companies.

The 1970s were a landmark decade for the capitalist development of the Brazilian countryside, with the release of government resources for production and the creation of important programs for the different Brazilian regions. In this decade, the creation of the National Alcohol Program (PROÀLCOOL) encouraged the expansion of sugarcane plantations in the country, by increasing the need for raw materials for the production of alternative fuels.

At the end of the 1970s, agriculture in Brazil was quite different from that of the 1960s, with a high degree of intersectoral integration at national level (MÜLLER, 1989).

However, the intense growth that took place between the 1950s and 1970s did not continue into the 1980s, as the global economic crisis interrupted the release of large government funds to finance modern agriculture (DELGADO, 1985).

The process of modernizing the technical base of agriculture slowed down in the 1980s (the so-called "lost decade"). The area under cultivation grew much more slowly than in the 1970s, stabilizing at 15% of cultivated areas. But despite the slower pace, agriculture continued to grow, unlike some countries that were also facing the economic crisis of the 1980s (GRAZIANO DA SILVA, 1996).

1.2. State stimulus for the modernization of agriculture

The Brazilian state played a key role in making the process of modernizing agriculture viable in the face of the country's industrialization and urbanization, enabling the sector to supply agro-industries with raw materials and expand its capacity to supply urban centers with foodstuffs.

Huge investments were made in importing the Green Revolution technological package and in developing agronomic research (the Brazilian Agricultural Research Corporation - EMBRAPA - was created in 1973), and credit lines were established to make the process of modernizing agriculture viable.

Regarding the state's participation in financing the modernization of agriculture in Brazil, Hespanhol (2008b, p. 375) states:

> The period between the end of the Second World War and the beginning of the 1970s was marked by the great expansion of the world economy and a significant role for the state in coordinating, regulating and stimulating productive activities.

The National Rural Credit System (SNCR)[2] was set up in 1965 with the aim of stimulating the incorporation of new technologies into agricultural activities.

However, according to Coelho (2001), the SNCR was clearly selective, as it favored only a few areas and producers, concentrating credit, mainly in the South and Southeast regions.

Hespanhol (2008b, p.382) reinforces this statement by pointing out that:

> Official rural credit, the main instrument used to promote the modernization of agriculture, was highly selective, as its offer was restricted to medium and large rural producers. Small tenants, partners and sharecroppers with little or no assets had no access to official credit because they didn't have the guarantees required by the financial system.

In addition to the SNCR, the agricultural policy of guaranteeing minimum prices, which stabilized prices, and the agricultural insurance policy, which compensated producers in the event of crop damage, were also important. However, these policies were discriminatory, as they only favored medium and large landowners. The fact is that the policies to stimulate modernization did not affect small food-producing properties (GRAZIANO DA SILVA, 1981).

Among the programs created by the government, we highlight the direct subsidies to agricultural activities, aimed at making inputs cheaper by paying part of the price of the

---

2 The National Rural Credit System (SNCR) was created by law 4829 of November 5, 1965, with the National Monetary Council (CMN) being responsible for regulating and the Central Bank for controlling its actions. The Central Bank, Banco do Brasil, Banco Crédito da Amazônia, Banco do Nordeste and Banco Nacional de Crédito Cooperativo were part of the SNCR (CASTRO, 2008, p. 2).

industrial product with state resources, as well as tax exemptions.

The 1970s also saw the creation of programs benefiting certain regions and activities, with concentrating and excluding effects. Among the main programs, in addition to the National Alcohol Program (PROÀLCOOL) mentioned in the previous section, the National Storage Program (PRONAGEM), the Northeast Integrated Areas Development Program (POLONORDESTE), and the National Livestock Development Program (PROPEC) stand out, Amazon Development Program (POLOAMAZÔNIA), Land Redistribution and Agribusiness Stimulus Program for the North and Northeast (PROTERRA), Pantanal Development Program (PRODEPAN) and the Cerrado Development Program (POLOCENTRO).

In this context, the creation of PROÀLCOOL in 1975 was intended to stimulate the production of alcohol in order to reduce the use of gasoline (GUANZIROLI, 2006).

The resources coming from the rural credit policy contributed to consolidating the relationship between rural and urban areas, with a notable strengthening of agro-industrial production and, consequently, of the country's economy. However, in the 1980s, the situation was not the same as in the previous two decades (1960s and 1970s), as there was a sharp reduction in the resources earmarked for financing agricultural activities, which hurt the industries producing machinery, agricultural implements, chemical fertilizers and biocides (DELGADO, 1985).

The areas where the agricultural frontier was expanding benefited greatly from the credit policy. The Cerrado regions, in the center of the country, were occupied with a view to increasing agricultural production in areas with slightly undulating terrain and easy mechanization. The acidity and low natural fertility of the soils were corrected with the new inputs and planting techniques inherent in the Green Revolution.

> [...] the state, above all through its planning action, centralizing decisions and determining the pace and direction of capital expansion, ended up becoming a kind of great driver of national development (GONÇALVES NETO, 1997, p. 141-142).

Rural credit expanded greatly in the 1960s and 1970s and fell sharply from 1980 onwards. Due to the high interest rates, large producers sought financing through export advances, advance purchases and loans from processing industries and cooperatives (GUANZIROLI, 2006).

Even with these changes in credit policy between the 1980s and 1990s, with a significant reduction in its volume, agricultural production continued to grow, but at a slower pace. For Guanziroli (2006), this resulted from the maturing of the sector's production matrix, thanks to the years of modernization driven by the state and the special credit programs created for specific sectors, such as cocoa and sugar.

In the 1990s there was a recovery in agricultural credit policy, although it was less than in the 1970s. The novelty was the inclusion of small producers in a more significant way, through the creation of the National Program for Strengthening Family Farming (PRONAF) in 1996. However, inequality continued in the volume of credit made available to family and corporate farming.

According to Santana (2005), Brazil continues to show great inequality in access to agricultural technology between regions and between farms. In 1995, according to Santana, only 16.7% of family farms received technical assistance, while 43.5% of large farms did. In terms of regions, the author shows that in the South, 47.2% of family farms received technical assistance.

received technical assistance and, in the North, only 2.7%. These figures, however, predate the implementation of PRONAF.

With regard to the use of industrial inputs, Oliveira (2003) points out the big difference between small and large farms in the 1990s. According to the author, 72.4% of large farms used tractors and only 8.5% of small farms did so.

On the subject, Hespanhol (2007, p. 274) states:

> The vast majority of farmers, especially small landowners, tenants, partners and sharecroppers, whose conditions of access to land were precarious, were not assisted by official rural credit and had greater difficulties in changing the technical basis of production and remaining in the countryside.

Despite the differences in the use of inputs and technologies between small and large farms and between Brazilian regions, Hespanhol (2008a) states that in the 1990s there were important changes in the way the countryside was understood in Brazil, and the new public policies implemented also benefited family farming.

In the 2000s, according to Guanziroli (2006), the government stopped using traditional production financing mechanisms and created new agricultural policies, as well as strengthening credit lines for small producers, such as PRONAF, created in 1996 by the Fernando Henrique Cardoso government and strengthened by the Luis Inâcio Lula da Silva government from 2003 onwards. In this decade, annual financing for family farming exceeded R$20 billion.

In 2003, under the government of Luiz Inâcio Lula da Silva, the Secretariat for Territorial Development was created, with the main objective of stimulating and coordinating development projects for rural territories. The Ministry of Agrarian Development (MDA) led the policies aimed at rural settlements and family farming, and the Ministry of Agriculture, Livestock and Supply (MAPA) led the policies aimed at medium and large producers. That decade also saw the creation of the so-called rural territories, starting in 2004, and the citizenship territories, starting in 2008, and the decentralization of public policies.

According to Hespanhol (2008b, p. 386):

> The excessive decentralization of public policies, despite the good intentions, may have adverse results than expected and reduce the level of democratization of decisions instead of increasing it, given the concentration of wealth and power in the hands of oligarchies in localities and regions of the country.

This shows how the Brazilian government treats family farming and agribusiness differently.

1.3. Formation of Agro-Industrial Complexes (CAIs) and the advance of agribusiness in the context of the modernization of agriculture in Brazil

The process of modernizing agriculture required greater investment in the agricultural production system, contributing to the strengthening of corporate agriculture in Brazil, with the formation of the Agro-Industrial Complex (CAI).

According to Müller (1989, p.45):

> The agro-industrial complex, CAI, can be defined, in formal terms, as a set formed by the succession of activities linked to the production and processing of agricultural and forestry products. Activities such as: the generation of these products, their processing and the production of capital goods and industrial inputs for agricultural activities; also: the collection, storage, transportation and distribution of industrial and agricultural products; and even more: financing, research and technology and technical assistance.

The rapprochement between agriculture and industry required greater participation by the state as a financier. This relationship took place through the contribution of state capital, rural producers and large national and international economic groups.

The formation of the CAI directed the country's agricultural production, with upstream industry supplying capital goods and inputs for agriculture, while downstream industry processed agricultural raw materials, i.e. agro-industry. This process is referred to by

Sorj (1986) as the "industrialization of agriculture".

For Marafon (1998), the formation of Agro-Industrial Complexes must be based on two concepts: the first is based on the concepts of *agribusiness* and *filière* and the second is the transition from the Rural Complex to the Agro-Industrial Complex.

In the first conception, the Agro-Industrial Complex takes place in a specific economic space from which interdependent activities are isolated, forming a productive set of systems (*agribusiness*) or chains (*filière*). This is known as the Macro Agroindustrial Complex (MÜLLER, 1989).

In the second concept, the formation of Agro-Industrial Complexes is understood as being based on the transformations in Brazilian agriculture from the so-called Rural Complexes of Colonial Brazil to the current changes, with the Agro-Industrial Complex. In this sense, according to Kageama (1987), the so-called Agro-Industrial Micro-Complexes are formed.

Because the formation of the agro-industrial complexes (CAIs) occurred through the centralization of industrial capital and the merging of agricultural production with industrial production, there was an important reorganization of agricultural production and the integration of capital (DELGADO, 1985), requiring greater production for the domestic market in order to meet the industrial demand for raw materials. "Therefore, the creation of the CAIs represented the start of the process of industrialization of the countryside and, consequently, of the productive restructuring of capital in Brazil" (MOTA and PESSÔA, 2009, p. 9).

As modernization expanded in Brazil, new crops were territorialized, especially those receiving incentives, i.e. those destined for export and those used as raw materials in agro-industries (GRAZIANO DA SILVA, 1981). Thus, the environment in the Brazilian countryside favored the formation of agro-industrial complexes, since agro-industry "reveals itself to be a support for land concentration" (DELGADO, 1985, p.181) and the valorization of capital takes place through the control of rural properties.

On this subject, Mota and Pessôa (2009, p. 2) state:

> From the 1970s onwards, Brazilian agriculture was no longer studied in terms of its rural complexes, but in terms of its agro-industrial complexes. In practice, this means a shift from agriculture strictly focused on local food production to a more intimate relationship between agriculture and industry, whereby industry provides the necessary elements for production and agriculture supplies raw materials for industrialization.

As agro-industries grew as processors of agricultural products and modernized, they began to demand more from agricultural production and from the state as a mediator and responsible for the infrastructure needed for better circulation of production. There were therefore greater incentives for the expansion of monocultures, such as soy and sugar cane.

In the case of sugar cane, Silva (1983, p. 52) states that:

> This integration has some extreme examples where production and equipment manufacturing come together to form real sugarcane conglomerates. The capital goods industry for sugar and alcohol production was highly concentrated in the 1970s and the largest equipment manufacturers also owned mills and distilleries that were among the top in the country.

According to Oliveira (2010), sugar-alcohol production was one of the industries that received the most incentives from the government, mainly due to PROÀLCOOL. In a short space of time, it received large sums of money, which "beans and basic foodstuffs have been demanding for a long time without getting anything" (OLIVEIRA, 2010, p. 22).

Since the 1980s, the term "agribusiness" has been used in Brazil, associated with the term Agroindustrial Complex. According to Mota and Pessôa (2009), this is the contemporary version of capitalism in the countryside, where production is organized using modern techniques and equipment, little labour and usually on large tracts of land.

Although the term "agribusiness" is recent, the practice has its origins in the *plantation* system, when large properties were used for monoculture production for export (FERNANDES, 2004).

This model received significant government investment and expanded considerably in the country, contributing to the generation of surpluses in Brazil's trade balance. The same did not happen with the production of foodstuffs for domestic consumption, as Hespanhol (2008a, p.191) points out:

> In Brazil, the discouragement of basic food production has been evident in recent decades, at the same time as the production of "important" agricultural *commodities* has expanded, especially soybeans, sugar cane, oranges and corn. Some traditional products such as coffee and cotton, which absorbed a large amount of labor in the countryside, have also lost their importance.

However, the 1980s, as we have seen, saw a reduction in government investment in the entire agricultural sector, due to economic instability. However, there was no decrease in grain production and, despite a slowdown in the occupation of Cerrado areas, there was an increase in productivity, made possible by the expansion of corporate agriculture.

The 1990s saw a resumption of investment, with a new restructuring of agricultural policy instruments, channeling more resources into rural credit, improving the infrastructure for marketing production and investing in new agricultural research, favoring the growth of agribusiness in the country (COELHO, 2001).

With the advance of the globalization process, agribusiness has organized itself to serve the international market. With this model, production is no longer seen only for the domestic market, but to maximize profits by exporting the products generated.

The advance of agribusiness in the Brazilian countryside, according to Coelho (2001), has established a model of private capital domination over producers, the economy and the territory, as corporations control production and monopolize exports.

According to Cleps Júnior (2010), there is a pact between the state and agribusiness in Brazil, demanding greater investment in transportation infrastructure and policies to support production. This is evident in the public policies developed in recent decades.

Agribusiness has become synonymous with production and, according to Oliveira (2006), the current productive conjuncture is linked to the globalization of capital and Brazil's modern countryside transforms agriculture into a profitable business regulated by profit and the international market. However, it leaves the country vulnerable in terms of its food sovereignty, since agribusiness is not aimed at producing food.

According to Teubal (2008), the main difference between the agribusiness model and the family farming system is that the former produces *commodities* for the international market, usually monocultures on large properties, and the latter features productive diversity on small properties, usually focused on foodstuffs to supply the domestic market.

In this sense, Oliveira (2006) points out that most of the food consumed in the country is produced on small farms, generating more jobs in the countryside, even with less investment than agribusiness.

With greater support from the state, agribusiness has increasingly expanded into Brazilian territory, competing for space with food crops, exploiting the workforce and manipulating its production (OLIVEIRA, 2006). In this context, the expansion of sugar cane into regions producing grains and foodstuffs, such as the Ivinhema river basin in Mato Grosso do Sul, our research area, has stood out in recent years.

Companies set up in strategic areas and territorialize themselves by buying land, leasing it or signing contracts with landowners, through which they start producing what the companies want, with the quality they demand. This was evident in the fieldwork we carried out on rural properties located in the Ivinhema river basin in Mato Grosso do Sul.

With the advance of agribusiness, driven by global capitalist logic, the use of modern technologies is increasingly intensifying which, according to Santos and Silveira (2001), are modifying daily life in the countryside and working relationships, with greater use of biotechnology, genetic engineering and microelectronic instruments, creating a new notion of productive time and space in the use of the land.

Agribusiness is playing an increasing role in Brazilian exports, showing considerable growth between the 1990s and 2000s, with the agricultural sector surpassing the livestock sector in terms of total income, according to the IBGE (Graph 1).

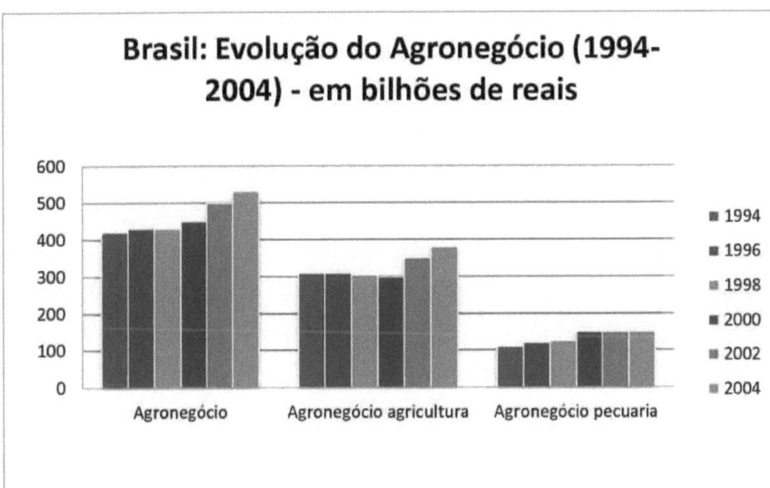

Graph 1: Brazil: Evolution of agribusiness (1994-2004) - in billions of reais Source: IBGE - Organization: TEIXEIRA, J. C. - November 2013

One of the most important products on the Brazilian agribusiness export agenda is sugar. Brazil is one of the world's largest sugar exporters and, in recent years, exports of anhydrous alcohol have also grown. In this context, the state of São Paulo stands out as the largest producer of this product nationwide and production is increasingly expanding towards the Midwest.

Despite being the target of criticism, modern agriculture has expanded the production of *commodities* and has been expanding in recent years, with a large concentration of production in the Center-South of the country and occupying the Cerrado areas of the Center-West, Northeast and North. As a result, sugar cane stands out in the production scenario, with large production in São Paulo, Paraná, Minas Gerais, Goias and Mato Grosso do Sul, in the latter with emphasis on the Ivinhema river basin, located in the southeastern part of the state, our study area.

1.4. The expansion of the sugar-alcohol sector in the context of the modernization of agriculture in Brazil: from PROÀLCOOL to PROETANOL.

The National Alcohol Program (PROÀLCOOL) was created under the government of Ernesto Geisel, through Decree Law No. 76.593 of November 14, 1975, with the aim of implementing the use of alcohol as a fuel in the face of the global oil crisis. To do this, it was necessary to expand its raw material, i.e. sugar cane plantations in Brazil.

According to Carvalho and Carrijo (2007, p. 3):

> Among the factors that contributed to the creation of PROÀLCOOL, it is worth highlighting the first "oil shock", which occurred in 1973, due to a substantial rise in oil prices as a result of the actions of OPEC (Organization of Petroleum Exporting Countries), made up mainly of the oil-producing countries of the Middle East. Current crude oil prices on the world market rose from 1.9 US$/barrel in 1972 to 11.2 US$/barrel in 1974, an increase of more than 550%.

In the first three years of the program, there was great incentive to increase the production of fuel alcohol, with the installation of distilleries attached to the sugar mills. The oil price shock of 1973 led the Brazilian government to look for new alternatives to alleviate the problem caused by this global event. The fall in the sugar market during this period, in a way, favored the production of alcohol, since the installed capacity for sugar production was used to produce alcohol (VEIGA FILHO and RAMOS, 2006).

The initial goal of the program was to produce three million liters of alcohol by 1980. According to Carvalho and Carrijo (2007), this goal was surpassed in 1979.

During this period, there was an increase in investment by traditional mill owners, both in distilleries attached to existing mills and in new distilleries. In addition, new economic groups set up other distilleries, mainly from land holdings. During that period, there was a 29% increase in the area harvested with sugar cane in Brazil, and in the state of Sao Paulo it was over 60% between 1975 and 1980 (MAGALHAES, *et. al.,* 1991).

In 1977, the Alcohol and Sugar Institute (IAA) approved resolution 01/77, classifying alcohol as follows:

- industrial refined alcohol (for the chemical industry);
- anhydrous alcohol (to mix with gasoline);
- fuel alcohol (hydrated - automotive).

The first alcohol-powered cars were introduced in 1978, after changes were made to the vehicle's engine (ANDRADE *et.al.*, 2009).

The new production techniques implemented as part of the modernization of agriculture were largely incorporated into sugarcane plantations, especially in the state of São Paulo, favoring their expansion and yields (SILVA, 1983). The 1970s presented favorable conditions for modernizing the sector, as the Brazilian government granted low-interest credit and implemented important agricultural programs.

Silva (1983) states that in 1973, Brazil exported almost three million tons of sugar, with revenues of close to 1.4 billion dollars, representing almost 18% of Brazilian exports. However, at the end of the following year sugar began its downward trajectory on the foreign market and, at the end of 1976, the foreign exchange revenue generated by the product stood at around 360 million dollars.

According to Michellon *et. al.* (2008), the first phase of this program began in that decade and lasted from 1975 to 1979:

> Alcohol production began to gain momentum with the help of the government, which granted financing and subsidies to the sector, and Petrobras was in charge of buying, transporting, storing, distributing and blending alcohol with gasoline. The government was also in charge of determining the selling price of the product. During this period, alcohol production rose from 555.6 thousand m^3 in the 1975/76 harvest to 2,490.6 thousand m3 in the 1978/79 harvest, an increase of more than 300% over the period (MICHELLON *et. al.*, 2008, p. 4).

Parallel to the fall in the sugar market, there was an increase in oil prices, which strengthened the production of alcohol. "The sugarcane production system ended up bringing together the widest range of necessary conditions - technical, economic and political - and an alcohol park of considerable size was built on it" (SILVA, 1983, p. 43).

The great advance of the sugar-alcohol sector in the Southeast and its imminent saturation meant that sugarcane cultivation expanded into other areas in the 1970s. This led to the expansion of production in Paranà, the Sâo Francisco Valley and Mato Grosso do Sul.

The rapid growth in Brazilian sugar cane production, which went from 76,610 thousand tons in 1968 to 153,858 thousand tons in 1981, an average of 5.6% a year, is particularly noteworthy when you consider the already large sugar cane areas at the start of the 70s. Other crops, such as soybeans and wheat in the South, recorded similarly expressive figures, but on a smaller scale. From 1975 onwards, however, growth took on extraordinary proportions, reaching 16.5% between 1976 and 1977 (SILVA, 1983, p.43-44).

Despite the expansion of sugarcane plantations, Silva (1983) draws attention to the decrease in the number of properties producing sugarcane between 1970 and 1980. This can be explained by the concentration of this production on medium and large properties, since sugar cane production is not viable for small producers.

Silva (1983, p. 50) points out:

> It would also be interesting to note that the decrease from 9,155 to 6,978 in the number of establishments growing sugar cane between 1970 and 1975 complements the characterization of the movement towards the upper strata. Productivity gains, supposedly derived from technological modernization, would be a way of making large-scale production viable.

In 1979, there was another global oil crisis, and the price of this product rose significantly. In view of this fact, the government gave greater encouragement to increasing the production of fuel alcohol, as it was expected that gasoline would increasingly be replaced by this product and, with this, the second phase of PROÀLCOOL began (MICHELLON *et. al.*, 2008).

On this issue, Veiga Filho and Ramos (2006, p. 50) state:

> The heyday, whose main motivation was the second oil shock in early 1979, took place between 1981 and 1985, with the establishment of many new distilleries, mainly in regions that were marked by the presence of large estates with extensive beef cattle in the state of São Paulo, but also in other states.

Thus, this period was marked by a considerable increase in the use of alcohol-fuelled vehicles in Brazil and the program reached its peak, but the sector suffered from the failure to increase productivity in the face of reduced investment in technology. The crisis of the 1980s, a *priori*, did not bring any major structural changes to the sector, which only really felt its effect in the second half of that decade.

According to Borges *et. al.* (1988) alcohol production reached 11.8 billion liters per year between 1985 and 1986. The increase in the circulation of alcohol-only vehicles further boosted the production of this product.

According to Ribas (1987) *apud* Carvalho and Carrijo (2007, p.3):

> The share of gasoline in liquid fuel consumption fell from 98.9% to 42.8% between 1975 and 1986 and the share of alcohol in that period went from 1.1% to 55.5%. In 1979, the percentage of alcohol vehicles rose from 0.5% to 66.2% in 1986, while the percentage of gasoline cars fell from 89% to 20.9%. Alcohol consumption in the automotive sector in 1984-86 was 89.6% of the total, with the chemical industry making up 3.5%, exports 3.1% and 3.3% for other purposes.

In 1986, according to Michellon *et. al.* (2008), the third phase of PROÀLCOOL began and lasted until 2003. In general, this period was marked by a crisis in the sector, with subsidies being cut and the suspension of funding to expand production, but with oscillations.

In 1989, for example, with the increase in the consumption of fuel alcohol, combined with other factors, there was a shortage of the product at the filling stations and, according to Veiga Filho and Ramos (2006, p. 48) "during this period, PROÀLCOOL reached its peak, with the structuring of a network of public tax and financial incentives, covering everything from ethanol producers to end consumers".

The Southeast region was the hegemon in PROÀLCOOL investments in the last decades of the last century, more specifically in the state of São Paulo, which saw more sugarcane expansion than the other states. According to Magalhâes *et al.* (1991), the Southeast received more than 60% of PROÀLCOOL investments until 1984. However, even though investments were lower in the other regions, they were still important.

In the 1990s, with the fall in oil prices, the sugar-alcohol sector was deregulated and turned more to sugar production, also due to the increase in the price of this product on the international market.

So, even though alcohol production was no longer so attractive, sugarcane production continued and, in the late 1990s and early 2000s, there was the crisis of overproduction, due to the decrease in the use of ethanol.

> During the 1990s, Brazil greatly expanded its sugar exports, largely as a result of the effects of the break-up of the Soviet Union and, consequently, of its preferential market, which was mainly supplied by Cuban production. This contributed to maintaining and even reinforcing the well-known instability of sugar prices on the free world market. On the other hand, after the aforementioned peak in August 1990, oil prices fluctuated less widely until the beginning of 1999, but remained at a level well below that of the first half of the 1980s (VEIGA FILHO and

RAMOS, 2006, p. 51).

The same authors point out that in the early 2000s, PROÀLCOOL was renewed:

> [...] mainly through corporate actions, articulating more and more economic, social and political segments, marked by the release of prices for sectoral products, the introduction of *flexfuel* vehicles, the possibility of an increase in ethanol exports and high prices, in the short and medium term, for oil on the world market (VEIGA FILHO and RAMOS, 2006, p. 48).

For Michellon *et. al.* (2008), the fourth phase of PROÀLCOOL began in 2003, after the rise and fall of the program, the increase in oil prices on the international market, the emergence of *flexfuel* vehicles and the search for alternative, less polluting energy. These factors contributed to the increased interest in using ethanol as a fuel. The program was renamed the National Ethanol Program (PROETANOL).

The introduction of *flexfuel* vehicles in 2003 undoubtedly contributed to the strengthening of sugarcane production in the country, given the need for greater production of raw materials for ethanol. According to Veiga Filho and Ramos (2006), sales of these vehicles amounted to 55% of light vehicles sold on the country's domestic market in 2005. There was also a significant increase in ethanol exports, which rose from 600 million to 1.4 billion liters between 2003 and 2004 alone.

The 2000s therefore marked an increase in ethanol consumption, given the search for alternative energy sources, preferably less polluting ones. As a result, new incentives for the installation of new plants were created by the Brazilian government and these companies looked for other areas in which to carry out their activities. The Midwest received some of these plants.

In the mid-2000s, Brazil was the leading producer of biofuels from sugar cane and, according to Carvalho and Carrijo (2007), this was made possible by the technological input developed for this production with the support of PROÀLCOOL resources.

These authors state:

> The country has already managed to replace 45% of the gasoline market with biofuel, while in the USA, the richest country in the world, not even 3% of consumption has been replaced. According to the Department of Sugarcane and Agroenergy, a comparison of the area harvested (in millions of hectares) and Brazilian sugarcane production (in millions of tons) between 1975 and 2006 shows an increase of approximately 211% and 112%, respectively (CARVALHO and CARRIJO, 2007, p. 10).

Table 1 shows the steady increase in the production of sugar cane and its derivatives in Brazil between the 2000/01 and 2009/10 agricultural years.

Table 1: Brazil: Sugarcane, Sugar and Ethanol Production (2000/01 to 2009/10).

YEAR- CROP	ANHYDROXED ETHANOL (m^3)	HYDRATED ethanol (m^3)	TOTAL ETHANOL (m^3)	SUGAR (tons)	SUGAR CANE (tons)
2000/2001	5.584.730	4.932.805	10.517.535	16.020.340	254.921.721
2001/2002	6.479.187	4.988.608	11.467.795	18.994.363	292.329.141
2002/2003	7.009.063	5.476.363	12.485.426	22.381.336	316.121.750
2003/2004	8.767.898	5.872.025	14.639.923	24.944.434	357.110.883
2004/2005	8.172.488	7.035.421	15.207.909	26.632.074	381.447.102
2005/2006	7.663.245	8.144.939	15.808.184	26.214.391	382.482.002
2006/2007	8.078.306	9.861.122	17.939.428	30.735.077	428.816.921
2007/2008	8.464.520	13.981.459	22.445.979	31.297.619	495.843.192
2008/2009	9.630.481	18.050.758	27.681.239	31.506.859	572.738.489
2009/2010	6.937.770	18.800.905	25.738.675	33.033.479	603.056.367

Source: Ministry of Agriculture, Livestock and Supply (MAPA)

Organization: TEIXEIRA, J. C. - November 2013

In view of what we have said so far, we can point to the following factors that contributed to boosting fuel alcohol production once again in the 2000s: the fluctuations and increase in oil prices in the face of tensions in the Middle East in that decade; the search for renewable energy sources; Brazil's mastery of the technology for producing alcohol from sugar cane; the use of its by-products; and the increase in domestic consumption with the greater production of bi-fuel vehicles.

On this subject, Hespanhol (2008a, p. 193) states:

> In recent years, the rise in oil prices and the growing use of ethanol in developed countries with the aim of reducing greenhouse gas emissions have contributed to the revitalization of the sugar and alcohol sector in Brazil and to the withdrawal of PROÀLCOOL from its dormant state.

The 2000s, therefore, were marked by an acceleration in the sector's growth and the consequent emergence of new plants in areas of sugar cane expansion, such as the Ivinhema river basin in Mato Grosso do Sul.

Silva (2008) points out that the production of ethanol as an energy source led to the migration of international capital, both to set up plants and to buy land, in the 2000s, accelerating production, especially in its second half.

Although the sugar-alcohol sector is seeking technological innovations, harvesting in many cases continued to be done by hand for much of the 2000s, raising questions about the exploitation of the workforce. Other issues were also raised for discussion in relation to the accelerated progress of the sector, mainly concerning the environment, due to the production and misuse of vinhoto, the soot from burning and the deforestation carried out to expand planted areas.

The space occupied by sugarcane plantations has undergone rapid changes in its natural and socio-economic aspects, driven by the power of capital, not taking into account environmental and social impacts, but rather the generation of income and profits.

Attention was also drawn to the replacement of important crops for the domestic market by sugar cane plantations. On this subject, Hespanhol (2008a, p. 192) states:

> The production of ethanol from corn in the US and sugar cane in Brazil and the cultivation of oilseeds to meet the demand for biodiesel in several countries also have negative effects on the supply of food for human consumption and, consequently, on the rise in its prices.

Mendonça (2008) points out that not long ago agrofuels were seen as alternatives to reduce the generation of carbon dioxide, making it possible to produce less polluting energy. However, the rapid expansion of sugarcane plantations is being blamed for the crisis in food production, for drastic changes in the environment, through increased deforestation, the removal of trees to facilitate mechanical harvesting, the burning still widely used in areas of crops and the aerial spraying of plantations, not respecting the limits of other nearby crops. This fact, which was quite significant in the 2000s, is still present in some areas, but to a lesser extent. This was verified in our study area during fieldwork.

The second half of the 2000s was marked by the so-called "*boom*" in the sugar-alcohol sector, when there were greater incentives for mills, decentralization of these companies and greater relocation to the Midwest and other parts of the country, as well as the incorporation of other companies in this sector. During this period, the region of the Ivinhema river basin, in Mato Grosso do Sul, stood out as a recipient of sugar and alcohol plants.

> [...] the search for sustainable, environmentally friendly and economically viable growth, in a context of growing concern about global warming, once again puts ethanol in a prominent

position (TONIN and TONIN, 2014, p. 62).

Tonin and Tonin (2014) point out that in the four decades since the implementation of Pro-Alcohol, the sugar-alcohol sector has experienced major fluctuations due to constant government intervention, but has always been influenced by changes in the international market.

However, since the early 1990s, with the extinction of the IAA, the gradual deregulation of the sector began, consolidated "with the publication of Law 12.490 of September 16, 2011, in which the National Agency for Petroleum, Natural Gas and Biofuels (ANP) began to regulate the biofuels sector" (TONIN and TONIN, 2014, p. 62).

In recent years, there has been an increase in competitiveness between companies in the sector and a reduction in government subsidies and incentives. Thus, after the sector's great advance in the second half of the 2000s, the 2010s began with signs of decline and crisis, which began with the global crisis of 2008-2009, even leading to the closure of some of these companies and the termination of some projects. Mato Grosso do Sul, for example, stopped receiving some companies planned for the beginning of this decade, as we will discuss in the final chapters of this work.

Monteiro (2010) points out that the deregulation of the sector and the reduction in state subsidies and incentives have led to changes in the pace of expansion and the spatial rearrangement of sugarcane agro-industries.

However, changes in the way the sugar-alcohol production process is carried out and increased competitiveness between companies have contributed to better structuring of the sector and its technological growth. Harvesting mechanization, for example, quickly replaced manual cutting and is present in most sugarcane fields, with a tendency to reach 100% of the area in a short time (SOUZA, 2014).

In addition, the sector began a process of diversification in sugar and ethanol production, targeting other alcohol, food and beverage markets (TONIN and TONIN, 2014), as a way of seeking new alternatives to face the crisis.

CHAPTER II: THE TRAJECTORY OF SUGAR CANE IN BRAZIL

In this chapter we analyze the history of the sugar-alcohol sector in Brazil and its economic repercussions, with the aim of understanding the current state of this sector and its expansion in the country, including in the Ivinhema river basin in Mato Grosso do Sul. Thus, we intend to deepen the discussion on how the sugar-alcohol sector expanded in the country and what factors influenced its expansion, as well as pointing out the current trends in the sector.

2.1. Historical aspects of the sugar-alcohol sector in Brazil

The history of sugar cane in Brazil began around 1530, when sugar played an important economic role in the European market. Its economic importance derives from the fact that it serves as a raw material in the manufacture of various products, such as sugar, rapadura, molasses, alcohol, brandy, feed, fertilizers, cellulose, etc., in other words, it was directly linked to the manufacturing sector. *In* addition, its juice, whether *fresh* or fermented, was a popular drink.

According to Segalla (1964) *apud* Szmrecsànyi (1979), the genus comprises five main species: wild cane (*saccharum spontaneum*), noble cane (*saccharum officinarum*), Chinese cane (*saccharum sinnensis*), Indian cane (*saccharum barberi*) and Javanese cane (*saccharum robustum*).

The most likely region of origin of sugar cane is the Asian continent[3] , from where it was taken to other parts of the planet and brought to Brazil by the Portuguese during the colonization period, specifically in 1533, when it was cultivated in the Captaincy of Sao Vicente (COSTA, 2008).

From 1535 onwards, the Northeast region became the center of Brazilian production, due to the quality of the alluvial soil, the vast coastal hydrographic network, the climate conducive to this crop and the greater proximity to Portugal, considering the place of shipment (FURTADO, 1961).

Initially, the plantations occupied the coastal regions of Pernambuco, Espirito Santo and Rio de Janeiro, later expanding to the São Paulo plateau. After the forest had been cut down, sugarcane plantations were cultivated for periods of up to 15 years, migrating to new areas of forest when the land became tired. The abandoned areas were then occupied by pastures

3 Authors differ on the origin of sugar cane. Most of them claim that it originated in Asia, probably in the south or southeast of that continent. However, some authors, such as Szmrecsanyi (1979) and Fauconnier and Bassereau (1970) point to New Guinea as the place of origin of this grass.

and, to a lesser extent, subsistence crops (CORDEIRO, 2008, p. 9).

Therefore, sugar cane cultivation in Brazil began at the beginning of colonization, when it played a role in promoting the settlement of the territory, keeping possible invaders away from the colony. The plantations were used to demarcate areas (FURTADO, 1961).

The Portuguese plan to make it possible to maintain possession of the land in an economically viable way was to create the first agricultural company in Brazilian lands, based on the experience of companies in the East and the Atlantic islands. Portugal could not have imagined how great the impact of this agricultural activity would be on the economy of the Old World. The sugar cane company that developed in Brazil guaranteed Portuguese permanence, as it set up a whole apparatus to make this monoculture viable, using slaves and providing technological development to produce more and more sugar (ANDRADE *et. al.*, 2009, p. 128).

Starting in the Northeast, this crop expanded rapidly using slave labor, with the aim of producing sugar, a product that was widely accepted on the European market. As a result, sugar soon became the mainstay of the colony's economy and its expansion was made possible by the Portuguese improving production techniques to reduce costs.

According to Andrade (1988), between 1530 and 1580, sugarcane plantations served to dominate the territory and led to the expulsion of the indigenous population from areas of fertile coastal soil. After this period, the expansion of these plantations became dependent on the exploitation of slave labor and the production of sugar in the Banguese sugar mills[4] . These mills dominated the sugar production process for almost 300 years, from 1580 to 1870.

According to Segatti (2009), between 1570 and 1610, the number of mills grew significantly, jumping from 60 to 400 production units.

The development of the sugarcane agro-industry played a major role in Brazil's economic history. For almost two centuries after the discovery, it was practically the only pillar on which the colonial economy was based (SZMRECSANYI, 1979, p. 43).

Thus, sugar exports drove the Colony's economy during the first centuries of exploitation. However, other activities were also important, mainly for subsistence, such as manioc, corn, rice, beans, as well as tobacco and cattle, which played an

---

4 The banana mills can be described as the first manufactures built in the Americas, because they were production units with a high social division of labor, both agricultural and industrial, and had the production of energy internalized in the production unit, which could be animal, in this case called trapiche mills, or hydraulic, when they were called real (WWF-BRASIL, 2008, p. 6).

important role in the sugar economy. Cattle were used for food, transportation and to power the mills. However, the most fertile land was used for sugar cane, which was detrimental to other crops (SEGATTI, 2009).

In the 18th century, with the changes in the mode of production and European mercantilism, sugar from other colonies was also dumped into international trade, damaging Brazilian exports. At that time, mining aroused greater economic interest, which lasted until the end of the gold cycle. With the end of this cycle, at the end of the 1780s, agriculture returned to prominence in the Colony, but sugar cane was no longer the only product of economic importance, given the advance of cotton production and coffee growing (PRADO JÛNIOR, 1987).

The 19th century was also not favorable to sugar production and, according to Segatti (2009, p. 25) "the country fell to fifth place on the list of sugar cane producers, accounting for 8% of world production, with production declining even more sharply at the end of the century".

According to Carone (1970) *apud* Szmrecsànyi (1979), the volume of sugar exports fell from 2,021,390 tons in the 1880s to 1,336,200 tons in the 1900s. This was due to exogenous factors (hegemony of beet sugar) and endogenous factors (high agricultural and industrial production costs, with the end of the slave system).

However, it was in this century that important changes occurred in the production system, with the introduction of the plow and changes in boilers in the industrial sector, which began to use bagasse as fuel. However, significant changes only really occurred with the shift in production hegemony from the sugar mills to the sugar mills in the 1930s, marked by the government's strong involvement in encouraging sugar to become more competitive on the international market, as it had lost ground to sugar beet production in European countries (COSTA, 2008).

The period from 1870 to 1930 was marked by political disputes between the mill owners, the owners of the mills and the owners of the banks. The intention was to turn the latter into sugar cane producers in order to supply the raw material to the central mills and processing plants.

The central mills were managed by foreign firms, usually English and French, and were encouraged by the government to increase production with a view to international trade, but they couldn't own land or slaves. As a result, they couldn't plant crops, a fact that justifies the attempt to make the owners of the Bangkoks mere producers and suppliers of raw materials.

For Andrade (1988), the transition from the banguês to the mills represented not only changes in production techniques, but also in labor relations, with slave labor being replaced by other relations, such as colonato, parceria, meaçâo, sujeiçâo, among others, which lasted until the 1960s, when salaried relationships were introduced and the figure of the "bóia fria" (cold buoy) worker emerged.

Costa (2008, p. 71), paraphrasing Andrade (1989), states:

> [...] the mill owners were the big winners in this dispute, because in six decades they managed, on the one hand, to eliminate the Central Mills from the dispute and the international firms abandoned their positions in sugar production and domestic trade and began to operate only in the international sugar trade. On the other hand, they reduced the power of the plantation owners by concentrating the land available for sugarcane plantations, maintaining in Brazil a structure that did not exist in other sugar-producing countries: the same player being an industrialist, landowner and slave owner at the same time.

Thus, we can see that the history of the sugar-alcohol sector in Brazil has always been marked by conflicts of interest, since its entry into the country during the colonization period, extending over the years and becoming more pronounced in the 20th and 21st centuries, as we will discuss in the next section.

2.2. The expansion of sugarcane plantations and the sugar-alcohol sector in the 20th and early 21st centuries in Brazil

In the second decade of the 20th century, Brazil resumed sugar exports because the First World War, which took place between 1914 and 1918, had damaged European production. However, according to Pina (1972), this didn't last long, because after this world confrontation, beet sugar production resumed in Europe and the world crisis of 1929 damaged Brazilian exports.

From the 1930s onwards, the sugarcane economy took a new direction, with the consolidation of the mills as the main unit of sugar production in Brazil, the end of the Central Mills and the owners of the banks becoming just suppliers of raw materials.

In that context, Costa (2008) points to some striking changes in the new phase of sugarcane production from the 20th century onwards, such as:

a) part of the mill owners became capitalist entrepreneurs with a different vision from the previous one, based on the landowning mentality;

b) greater state involvement in the production process and in regulation, culminating in the creation of the IAA in the 1930s;

c) changes in labor relations on the farms with the introduction of salaried work and the "cold buoys" and;

d) The expansion of sugar cane plantations to the Center South, especially the state of São Paulo, which became the leading producer of this product in 1946.

The creation of the IAA in 1933 gave a new direction to sugarcane production in Brazil, as its main goal was to regulate this agro-industry. In 1941, with the advent of the Sugarcane Plantation Statute (ELC), the IAA's functions were expanded and the institute also began to control transactions between mills and suppliers (COSTA, *et. al.*, 2010).

However, according to Szmrecsànyi (1979), the IAA acted in the interests of the oligarchies, managing the sugar and alcohol markets, including monitoring the areas planted with sugar cane.

The shift in sugarcane production to the Center South is explained by the fact that this region has the largest domestic market in the country, the fertile land allowing for greater productivity, the industrial park in the sector producing goods for the sugar-alcohol industry and the limited land in the Northeast.

During the Second World War (1939-1945), sugar exports to the East and Europe resumed, at satisfactory prices but in small quantities. At the beginning of the 1950s, Brazil exported only 15,821 60-kg bags of sugar. This scenario changed rapidly in the following years and by the end of the 1960s Brazil was exporting 18 million bags of sugar (PINA, 1972).

During the 1960s, a period of agricultural modernization in Brazil, government incentives for sugarcane production intensified due to the growth in sugar exports. In that decade, the Fund for the Consolidation and Promotion of the Sugarcane Agroindustry was created, aimed above all at encouraging trade in this product on the international market (BRAY *et. al.*, 2000).

The creation of the SNCR in 1965 further strengthened the sector and enabled the formation of the sugarcane agro-industrial complex through the release of credit at low interest rates. However, as we pointed out in the previous chapter, the injection of capital into the countryside, via the SNCR, contributed to greater concentration of land ownership, leaving small farmers on the margins of the process.

The period from 1969 to 1974, according to Xavier and Bombardi (2009), was marked by the significant capitalization of sugar mill owners and considerable changes in the Brazilian countryside, including the expansion of sugarcane production into new areas, including food production. During this period, according to Segatti (2009), the country increased its sugar exports by around 180%, reaching 2nd place in the world.

In the 1970s, the sugar-alcohol sector relied on new agricultural implements in the state of São Paulo, mainly harvesting machines. According to Camargo *et. al.* (2008), the use of machines to harvest sugar cane made production approximately 30% cheaper and, on the other hand, the European market preferred to buy alcohol from "clean" sources, i.e. without the use of burning.

According to Silva (1983), 1974 was a favorable year for sugar exports, with revenues reaching close to 1.4 billion dollars, almost 18% of the Brazilian total. However, by 1976 the figure had fallen to 360 million dollars, due to the government's greater emphasis on alcohol production and increased international competition.

In this sense, Souza (2011, p. 86) points out:

> The context of the 1970s, due to factors such as the oil crisis caused by conflicts in the Middle East, the creation of the Organization of Petroleum Exporting Countries (OPEC) and the fall in the price of sugar on the international market, can be "read" in this perspective, when the Brazilian government began to encourage the production of alcohol as an alternative to imported gasoline and oil as a source of energy. This led not only to the expansion of the activity, but also to a technical-scientific apparatus in the search for productivity and efficiency with research into the genetic improvement of sugar cane, thus consolidating the relationship between the production of knowledge, development and capital.

As we pointed out earlier, in the 1970s, the agro-industry made significant progress with the creation of PROÀLCOOL, government incentives and the closer relationship between the sugar-alcohol sector and the automobile industry, with the growth in the use of hydrated alcohol as an alternative to replace gasoline.

> In 1972, the installed capacity of the national alcohol industry was 4.6 million liters/day, 900,000 of which were located in the North/Northeast and 3.7 million in the Center-South. During the 1970/71 harvest, 164 distilleries operated, whose capacity in 180 days of effective work (harvest period) allowed for an annual production of 828 million liters. The highest production achieved until then had been 766 million liters in the 1967 calendar year (or 726 million in the 1966/67 harvest) (SZMRECSÀNYI, 1979, p. 308-309).

The same author points out that, at the end of the 1970s, there were 200 mills in operation in the country and suppliers numbered around 42,000.

For Baccarin (2005), this growth was in fact visible from 1975 onwards, when there was an acceleration in the development of sugar-alcohol agro-industries and a major expansion of sugarcane plantations, driven by the creation of PROÀLCOOL.

We should remember that this program came about because of the 1973 oil crisis and the need to develop an alternative fuel. The government therefore made credit lines available to finance the expansion of sugarcane plantations and obtain the raw materials needed for the mills, as well as encouraging the production of alcohol-powered vehicles. Control over the development of PROÀLCOOL was the responsibility of the National Alcohol Council (CNAL) and the National Alcohol Executive Commission (CENAL) (SILVA, 1983).

> This stimulated the entry of new entrepreneurs and the expansion of sugarcane plantations into regions with little or no tradition of sugarcane cultivation, such as the west of São Paulo, the north of Paraná, the Triângulo Mineiro and areas of the Midwest (COSTA *et.al.*, 2010, p. 4).

In the 1970s, this program achieved satisfactory results and the expansion of sugarcane plantations in the Center-South, especially in the state of São Paulo, occurred rapidly, requiring the search for new areas to expand this crop.

> The move to these regions, which were initially poorly equipped for planting sugarcane, represented a natural obstacle to achieving the official goals. However, before this slowed down the progress of the program, the lack of resources interrupted the approval of new projects from the second half of 1981. The beginning of the 1980s, unlike the decade that was ending, seemed to bring embarrassing prospects for the sugarcane sector (SILVA, 1983, p. 43).

According to the same author, Brazilian sugar cane production grew by more than 100% between 1968 and 1981, with an average increase of 5.6% per year.

Despite this significant growth, Brazil failed to take the lead in the production of this product, falling from 15th to 11th in the world *ranking of* sugar cane production between 1970 and 1980, according to data from the Food and Agriculture Organization of the United Nations (FAO), presented by Silva (1983). The leading producers during this period were Peru and Indonesia.

The use of alcohol-powered vehicles was not initially as successful as hoped, as they began to present a number of mechanical problems. The cost of maintenance did not compensate for the lower cost of fuel and sales of these vehicles fell in the following decade (1980s), when the sector went into crisis. The efforts of the Brazilian government and the automobile industry failed to contain the crisis and sales of these vehicles declined both domestically and abroad.

Despite the problems that occurred from the 1980s onwards, the period from 1975 to 1990 was marked by strong government investment in the sugar-alcohol sector, which led to the purchase of land and the setting up of attached and autonomous distilleries, as well as the modernization of production, with resources provided by the state. "These loans were provided at real interest rates below inflation, with a three-year grace period and 16 years to pay off the principal" (WWF-BRASIL, 2008, p. 8).

On this subject, Xavier and Bombardi (2009, p. 3), paraphrasing Bray *et. al.* (2000), state:

> Historically, and especially since Proàlcool (National Alcohol Program - 1975), there have been intense changes in the Brazilian countryside, with more and more land being set aside for sugar cane production. The actions of the State, together with major capital investments, brought about changes in the technical base with a view to boosting the sugar-alcohol agro-industry.

The sugarcane agro-industrial companies concentrated a greater volume of capital as a result of favourable government actions, making it possible to make greater investment in the expansion of the sector.

> The maintenance of the power, both political and economic, of the agrarian bourgeoisie at the helm of the sugarcane agro-industrial complex allowed the businessmen to invest more in sugarcane production, accounting for most of the raw material to the detriment of the suppliers, especially the small ones (OLIVEIRA, 2007, p. 4).

However, the IAA and PROALCOOL meant that the central role in organizing sugar-alcohol production remained with the state. Thus, from the creation of the IAA until the end of the 1980s, the government's interventionist position overpowered the actions of private companies. This situation only began to change with the end of the IAA and the PROALCOOL crisis in the 1990s.

In addition to the IAA, the government used other bodies in its interventionist policy, such as the Ministry of Finance, the Ministry of Industry and Trade, the Ministry of Mines and Energy, the Ministry of Agriculture and Agrarian Reform and Petrobras. According to Oliveira (2007, p. 4), the main functions of these bodies were:

- fixing the prices to be paid by the government to sugar cane, sugar and alcohol producers;

- the government sets fuel prices for retail outlets by means of a subsidy represented by the alcohol/petrol parity;

- maintaining strategic stocks of alcohol;

- determining financing for the installation of new production units or the re-equipping of existing ones; and

- financing alcohol stocks.

Between 1950 and 1980, the increase in oil prices favored growth in the production of alcohol fuel. In 1950, a barrel of oil cost US$1.75 and, at the beginning of the 1980s, this value reached US$26.00 (CARVALHO and SANTOS, 2012).

However, during the 1980s, there was a sharp drop in the price of this product, which affected alcohol production. It is worth remembering that, in this decade, there was a reduction in the granting of subsidized credit by the government, which affected production. According to Szmrecsànyi *et. al.* (2008), between 1986 and 1990, sugar-alcohol production stagnated, reducing the sale of alcohol-powered vehicles.

Despite the recession of the 1980s, Costa *et. al.* (2010) state that the 1990s was the worst phase the sugar-alcohol complex went through, with businessmen and political representatives mobilizing to create new public policies for the sector.

The 1990s were marked by strong criticism of PROÀLCOOL from some parts of the government and civil society. With the re-democratization of the country, a negative image was formed of this program, due to the fact that it had been created during the dictatorship, in addition to problems that arose during the program, such as: indebtedness of the mills, tax evasion, negative environmental impacts, child and slave labor in the sugarcane fields (COSTA, 2008).

Proponents of the sugarcane sector sought alternatives to improve their image in the eyes of the population and to obtain new investments from the government. An example of this was the creation of the Parliamentary Front of the Sugar and Alcohol Sector in the National Congress in 1996, made up of parliamentarians interested in defending the producing states (SOUZA, 2011).

However, the return on investment didn't come quickly, picking up at the end of the 1990s and increasing considerably in the 2000s. The government tried to encourage the expansion of agribusiness in general, but the advance of the sugar-alcohol sector required special monitoring by the government.

Despite the extinction of the IAA at the beginning of the 1990s, the sector now had the support of the Interministerial Sugar and Alcohol Council (CIMA) and the Sugarcane and Agroenergy Department (DCAA), linked to the Ministry of Agriculture, Livestock and Supply (MAPA).

According to Oliveira (2007), the creation of the Union of the Sugarcane Industry (UNICA), the Coalition of Sugar and Alcohol Producing Entities (CEPAAL), Brasil Alcool, the Bolsa Brasileira de Alcool S/A (BBA) and the Council of Sugarcane, Sugar and Alcohol Producers of São Paulo (COSECANA) were also important for the strengthening and self-affirmation of the sugar-alcohol sector.

At the end of the 1990s, several companies in the sugar-alcohol sector were in debt and the government launched a new form of subsidy, with financing through the Bank of Brazil, aimed at stockpiling 1.3 billion liters of anhydrous and hydrated alcohol (WWF-BRASIL, 2008).

Government incentives and the good acceptance of sugarcane products on the national and international markets led to the resumption of growth in this sector, especially since the mid-2000s, with new planting areas and the establishment of new mills and distilleries. The rise in oil prices has contributed to an increase in the production of fuel alcohol and, as a result, there has been an expansion in the areas of sugar cane cultivation.

In 2000, the price of a barrel of oil averaged US$ 30.00, in 2005 it reached US$ 54.00 and in 2007 it reached US$ 93.00. Ethanol exports have also grown considerably: in 2004 Brazil exported 500 million liters and in 2006 it exported 3.4 billion liters. In addition, domestic consumption of the product has grown considerably, given the introduction of bi-fuel vehicles in 2003 (SZMRECSANYI *et. al.*, 2008).

Table 2 shows the evolution of the sugar-alcohol sector in Brazil between 1980 and 2013. The evolution of sugar cane, sugar and alcohol production over this period was constant, with minor fluctuations. Thus, the data presented reaffirms the discussion made so far in this chapter.

The concentration of production took place in the Center-South, confirming it as the center of the country's sugar and alcohol production and reinforcing the reduction in the relative participation of the North-Northeast.

Table 2 shows that sugarcane production grew significantly over the period analyzed.

At the beginning of the 1980s, the country produced 123,681 thousand tons, rising to 223,178 thousand tons in the 1985-86 harvest, but this production remained almost unchanged until the end of the decade, considering that in the 1990-91 harvest, the total amount of sugarcane harvested was 222,430 thousand tons.

Table 2: Brazil: Sugarcane, sugar and ethanol production (1980/81, 1985/86, 2000/01, 2005/06, 2010/11, 2012/13 and 2014/15 harvests).

REGIONS	Sugarcane (thousand tons)	Sugar (thousand tons)	Ethanol (thousand m^3)		Total ethanol (thousand m3)
			Anhydrous	Hydrated	
VINTAGE: 1980-1981					
North-Northeast	40.248	3.001	283	368	651
Center-South	83.433	5.253	1.823	1.233	3.056
Brazil	123.681	8.254	2.106	1.601	3.707
VINTAGE: 1985-1986					
North-Northeast	59.804	3.199	383	1.647	2.030
Center-South	163.374	4.834	2.813	6.986	9.799
Brazil	223.178	8.033	3.196	8.633	11.829
HARVEST: 1990-1991					
North-Northeast	52.235	2.857	199	1.608	1.807
Center-South	170.195	4.509	1.088	8.620	9.708
Brazil	222.430	7.366	1.287	10.228	11.515
VINTAGE: 1995-1996					
North-Northeast	44.547	3.328	421	1.346	1.767
Center-South	204.383	10.185	2.588	8.256	10.844
Brazil	248.930	13.513	3.009	9.602	12.611
HARVEST: 2000-2001					
North-Northeast	49.719	3.554	818	709	1.527
Center-South	207.099	12.644	4.803	4.262	9.065
Brazil	256.818	16.198	5.621	4.971	10.592
HARVEST: 2005-2006					
North-Northeast	48.345	3.808	791	718	1.509
Center-South	336.784	22.015	7.276	7.036	14.312
Brazil	385.129	25.823	8.067	7.754	15.821
HARVEST: 2010-2011					
North-Northeast	63.464	4.505	910	1.082	1.992
Center-South	556.945	33.501	7.413	17.971	25.384
Brazil	620.409	38.006	8.323	19.053	27.376
HARVEST: 2012-2013					
North-Northeast	55.720	4.149	1.114	750	1.864
Center-South	532.758	34.097	8.730	12.632	21.362
Brazil	588.478	38.246	9.844	13.382	23.226
HARVEST: 2014-2015					
North-Northeast	59.381	3.633	1.455	8 55	2.310
Center-South	575.386	33.721	11.279	15.609	26.888
Brazil	634.767	37.354	12.734	16.464	29.198

Source: CONAB, UNICA, BIOSUL, UDOP and MAPA.

Organization: TEIXEIRA, J. C. - May 2015

In the 1990s, sugarcane production did not grow as much as expected. Table 2 shows that, in the 1995/96 harvest, the country produced 248,930 thousand tons and, at the beginning of 2000, 256,818 thousand tons, showing slow growth in that period. This reinforces what we said about the crisis in the sector, which began in the mid-1980s and continued until the end of the 1990s.

In the 2000s, the sector grew again and production reached 385,129 thousand tons in the 2005-06 harvest, reaching 620,409 thousand tons at the start of 2010, an increase of almost 100% over five years.

The 2010s began with sharp fluctuations in sugar cane production. In the 2012-13 harvest, there was a slight drop in sugar cane production in Brazil compared to the previous harvest, and the country recorded a total of 588,478 thousand tons, reflecting the crisis in the sector nationwide. However, production resumed its growth and exceeded 630,000 tons in the 2014/15 harvest, 90.6% of which was in the Center-South.

With regard to sugar production, table 2 shows that in the 1980s there was a significant drop, when the total produced fell from 8,254,000 tons in the 1980/81 harvest to 7,366,000 tons in the 1990/91 harvest. However, growth resumed during this period, reaching a total of 16,198 thousand tons in the 2000-01 harvest and 25,823 thousand tons in the 2005-06 harvest. In the 2010-11 harvest, 38,006 thousand tons of sugar were produced in Brazil, and this remained unchanged in subsequent harvests.

With regard to the ethanol produced during the period, in the 1980s we saw a considerable increase in production, from 3,707 thousand m^3 in the 1980/81 harvest to 11,515 thousand m3 in the 1990/91 harvest. However, the last decade of the 20th century was not favorable, as we described in the course of this text, and ethanol production fell to 10,592 thousand m3 in the 2000/01 harvest.

From the 21st century onwards, with the expansion of sugar cane cultivation and new investments in the sector, ethanol production resumed its growth and reached 27,376 thousand m3 in the 2010-11 harvest. However, the 2010s did not start optimistically and production fell to 23,226 thousand m3 in the 2012/13 harvest, but grew again in the 2014/15 harvest, reaching a total of 29,198 thousand m3. According to Souza (2014), this variation in national ethanol production reflects the fluctuations in the ethanol market in recent years.

According to the National Supply Company (CONAB), the first survey of the 2015/2016 sugarcane harvest indicates a growth of around 3% in production compared to the previous harvest, bringing production to over 650 million tons. The area occupied is expected to increase by 0.7%, i.e. the rate of growth has slowed down compared to what happened at the end of the 2000s. The CONAB study also points out that most of the sugar cane harvested will be used to produce ethanol (56.2%), with production expected to increase by 1.9%. Most of the ethanol produced will be used to meet the increase in the mixture with gasoline from 25% to 27%. Sugar production is also expected to grow by around 5% compared to the previous harvest.

2.3 The economic and socio-spatial effects of the advance of the sugar-alcohol sector in Brazil

In view of the growth in sugar cane cultivation areas in Brazil in recent years, questions have been raised about the possible environmental and socio-economic effects caused by this monoculture. One of the economic impacts pointed out is the replacement of food crop areas by sugarcane plantations.

However, the defenders of sugar cane claim that this expansion is taking place on degraded pastures and not on food crops. In any case, this monoculture promotes land concentration because it is cultivated on a large scale.

However, "the effects of sugarcane expansion on the country's land structure have not been properly assessed by most scholars of the process, partly due to the lack of recent and reliable data" (SZMRECSANYI *et. al.*, 2008, p. 52).

In recent years, mechanized production and the presence of industrial capital in sugarcane plantations have become increasingly common. The implementation of modern machinery and the use of biotechnology and genetic engineering makes production viable and demonstrates the interest of businessmen in maintaining the sector's growth. This has negative effects on work in the fields, as modernization redefines the production process and reduces the use of labor in the fields.

In this regard, Thomaz Jûnior (2002, p. 136) states:

> [...] technology itself is not against work, or against the worker, but it is in favor of capital, that is, the work of technical-scientific research takes place within the framework of the production/reproduction of capital, where discovery (machines, etc.) is the synthesis of capital accumulation itself.

In recent years, sugarcane monoculture has become the country's third largest crop, second only to corn and soybeans. In Mato Grosso do Sul, despite the predominance of those products, between 1995/96 and 2006, the area under sugar cane grew by 129%, according to data from the IBGE's Agricultural Census.

In the last decade of the 20th century and the beginning of the 21st century, the Brazilian state partially stopped regulating the sugarcane sector, when there was a reduction in exports of agribusiness products. According to Costa (2008), this was the result of the overvalued dollar and the sector's liquidity and debt crisis, generating profound changes in the sugarcane agro-industrial complex and a reduction in the number of mills in operation. At the same time, there was greater openness to transnational companies.

From 2002 onwards, however, with the recovery of sugar prices on the international market, the sugarcane agro-industrial complex gained new dynamism. During this period, there was an increase in domestic demand for ethanol, given the intensification of environmental discussions and the launch of bi-fuel cars in Brazil. This phase is reminiscent of the golden period of PROÀLCOOL, from 1975 to 1983 (WWF-BRASIL, 2008).

Some factors have favored the resumption of growth in the country's sugar and alcohol agribusiness since the beginning of the 21st century, such as the good acceptance of the products (sugar and alcohol) both on the domestic and foreign markets, the rise in international oil prices, the opening up of European trade to Brazilian sugar, concerns about global warming based mainly on the Kyoto protocol and the search for less polluting fuels, as well as the low costs of Brazilian production compared to countries that produce sugar from sugar beet and alcohol from corn.

> The arguments most often used in defense of agrofuels are based on the possible environmental gains generated by their production. In this sense, the mainstream media has played a fundamental role in unifying the environmentalist discourse with the new agribusiness ventures, especially with regard to the production of ethanol and biodiesel (XAVIER and BOMBARDI, 2009, p. 2).

The first decade of the 2000s saw constant growth in sugar cane cultivation, which accelerated in the second half, as we showed earlier. This production took place mainly in central southern Brazil, especially in the state of São Paulo, which in 2006 accounted for more than 60% of production.

> [...] plantations tended to concentrate in the state of São Paulo and expand from there to other states in the center-south of the country. On the other hand, the traditional sugarcane-growing states of the Northeast, mainly Pernambuco and Alagoas, drastically reduced their sugarcane production and cultivation area and, as a result, lost importance in the context of the sugar-alcohol sector (HESPANHOL, 2008, p. 22).

Thus, the second half of the 2000s was marked by an acceleration in sugarcane production to meet market demands, despite a slowdown at the end of the decade.

This crop expanded and became territorialized in new areas, accompanying the decentralization of the mills.

This has led to a new phase of sugarcane expansion, repeating what had already happened in other periods, such as the 1960s, with the increase in sugar exports, or since the implementation of Pro-Alcohol; in other words, a process of annexation of areas which in the past were characterized by the production of diversified crops and which today have been transformed into large sugarcane-producing estates (XAVIER and BOMBARDI, 2009, p. 12). 12).

In this context, Mendonça (2009) draws attention to the fact that the advance of sugar-alcohol agribusiness is taking place on traditional areas of food production, on areas of settlements and extractive communities, altering social relations, considering that for corporate-based agriculture what counts is the guarantee of profits.

Faced with this expansionist trend in the sector, Mato Grosso do Sul, especially in the Ivinhema river basin, received new mills and underwent significant changes in the regional agricultural space, with other crops and pastures being replaced by sugar cane plantations.

According to Szmrecsànyi (1979), the factors determining the location of industrial establishments are numerous and varied, with ease of transportation and proximity to raw materials standing out. However, nowadays the "government incentive" factor, such as tax exemptions and the availability of official credit, also plays an important role in this process. This aspect has been paramount in the spatial mobility of the sugar and alcohol industries.

Given the increase in the consumption of sugarcane derivatives, it is estimated that it will be increasingly necessary to incorporate new areas into the production of raw materials for ethanol and sugar, which opens up the question of food sovereignty. It is therefore up to the government to redirect agricultural policies so that the expansion of sugarcane monoculture does not jeopardize the production of basic foodstuffs for the population.

The growth in the international trade of ethanol is evidenced by the fact that the number of countries with which Brazil trades the product has increased. According to Xavier (2010), in 2006, the country exported ethanol to 39 countries and this number rose to 43 in 2007 and 58 in 2008. However, this market has slowed down since the global crisis of 2008-2009.

In this context, Petrobras played an important role, participating more intensively in this market, exporting more than 200 million liters of ethanol in 2007. A large part of the funds invested came from the National Bank for Economic and Social Development (BNDES), and in 2008 the sector had a credit portfolio of R$29 billion with the bank (BRITO, 2008).

In recent years, the mechanization of sugar cane cutting has increased significantly, reducing the need for manpower. Burning is on the decline, although it still occurs, even in areas that have recently been incorporated into sugar cane production. Another issue that deserves to be analyzed is the outsourcing of workers, which is increasingly common (SOUZA, 2014), but this aspect is outside the scope of this research.

On this subject, Souza (2011, p. 118-119), paraphrasing Alves (2000), states:

> The introduction of technological innovations opens up space for the flexibilization of production and work relations in companies, which has direct repercussions on the segmentation of the workforce and the social division of labor between companies. This is the case with the reduction in the number of workers with permanent contracts and the increase in the number of temporary workers.

As the sugarcane sector has advanced, programs have been set up in the country to improve plant quality, increase productivity and help producers fight pests.

On the subject, Goes and Marra (2009, p. 2) believe that:

> [...] the sustainability of sugarcane agricultural production in Brazil is based on the ability to respond to pests, diseases and climatic variations and this has been possible with the continuous supply of resistant varieties.

Graph 2 shows the variation in the area harvested with sugar cane between 1993 and 2014, according to IBGE data. During this period, we can see the constant growth of this crop, which justifies the increase in total production shown above. Even in the 1990s, considered a crisis period for the sector, expansion continued, albeit at a slower pace. We also saw a slowdown in the first few years of the 2010s, as a result of the crisis that began in the sector.

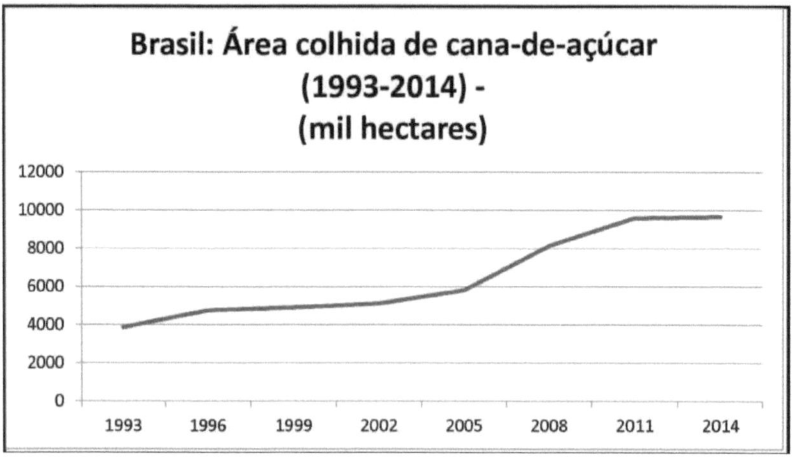

Graph 2: Brazil: Sugarcane harvested area (1993-2014) - (thousand hectares).

Source: IBGE - Municipal Agricultural Production.

Organization: TEIXEIRA, J. C. - March 2015.

Graph 2 shows the acceleration in sugarcane expansion that took place until the end of the 2000s, as demonstrated earlier.

On this expansion, Souza (2010, p. 7) points out:

> [...] with the creation of an international environment favorable to the transformation of ethanol into a *commodity*, added to the good prices of sugar on the international market, there are now prospects that point to an expansion in order to supply the national and possibly international demand for this agrofuel.

In 2009, the country had 35% of the world's ethanol production and was the largest exporter of sugar, with 350 mills installed in the country, controlled by 80 different groups (SEGATTI, 2009).

Currently, sugar cane occupies more than 9 million hectares throughout the country. Although the crop is grown in almost every state in the Federation, the Center-South remains the main production area, concentrating around 90% of production, according to CONAB data.

The expansion of sugarcane agribusiness has significantly changed Brazil's agricultural space, promoting greater land concentration and territorial disputes with food crops, as well as worsening the exploitation of the workforce and environmental impacts, as we have seen.

As a result, the Agroecological Zoning of Sugarcane (ZAE) was created in 2009 to organize the expansion and production of sugarcane in Brazil.

The creation of the ZAE is important in terms of regulating the expansion of sugarcane monoculture, since the ordering of this crop, based on capitalist interests, has been accompanied by socio-environmental impacts, requiring state intervention to regulate and discipline the use of agricultural land.

However, according to the Brazilian Agricultural Research Corporation (EMBRAPA)[5] , there is no need for sugar cane to occupy food production areas, considering the large tracts of unproductive land that exist in the country, as well as the degraded areas occupied by pastures that could be "better" utilized with this crop. But we know that, in a capitalist country, what drives the organization of production is the possibility of greater income and profit, so there is no widespread concern that other crops could be harmed. This is evident in the fact that sugarcane cultivation has expanded precisely in the region with the highest food production, the South Center of the country.

The aim of the ZAE is not to prevent the advance of sugarcane cultivation into areas of food production, but to prevent the advance of this crop into areas of environmental fragility, such as the Pantanal and the Amazon.

The area with the highest concentration of sugarcane plantations, in the South Central region of the country, is called by Thomaz Junior (2008) the "agribusiness polygon", where agribusiness is territorialized in privileged areas of important river basins, comprising part of the states of São Paulo, Mato Grosso do Sul, Mato Grosso, Goiás, Minas Gerais and Paranà.

In this perspective, Souza (2010) states that territorial disputes in this area will be inevitable and, increasingly, portions of land currently occupied by corn, wheat, beans, soybeans or pastures will be replaced by sugar cane, given current trends.

And this will happen regardless of the existence of millions of hectares that are vacant or underused by cattle ranching, but which don't have the locational advantages for the commercial and economic viability of sugarcane agribusiness. Bearing in mind that the only rationality that counts is capitalist economic rationality, the activity that presents the best possibilities for extracting land income from the same land will win this dispute for territory (SOUZA, 2010, p. 13).

Another important aspect is the growing number of agrarian conflicts in the areas of

5 Information available at: http://*www.embrapa.br*. Accessed on: Mar. 2013.

sugarcane expansion. Figure 1 shows an encampment of landless workers near the Aurora mill in the municipality of Anaurilândia, Mato Grosso do Sul.

Figure 1: Anaurilândia - MS: Landless workers' camp near Usina Aurora - MS 276.

Source: Field research - photo by the author - June 2013.

Faced with the advance of sugarcane cultivation and the modernization of the sector, social movements are "raising their banners" against the territorialization of agribusiness, the unemployment generated and the environmental impacts produced.

According to Oliveira (2008), the advance of sugar-alcohol agribusiness has no geographical limits and has even reached states in the Amazon, a worrying fact due to its environmental impacts and the threat to food sovereignty. According to the author, the expansion of biofuels will continue to lead to a reduction in food production.

The new configuration of the Brazilian countryside, driven by the expansion of sugarcane cultivation and the establishment of new processing units, is the result of the expansion of capital imposing its force and materializing through sugarcane agro-industrial capital.

In addition, the products of this new dynamic of productive spatial reordering signal changes in land rents around the mills, a reduction in the supply of beef on the domestic markets and, finally, a new onslaught by livestock on the forest through deforestation in order to regain the spatialization lost to sugarcane (FARIAS and SILVA, 2012, p. 2).

IBGE data presented by Farias and Silva (2012, p. 5) show the growth of this activity in the Amazon in the 2000s: "[...] in 2006 there was an expansion of 68.9% in

Tocantins, 55.1% in Amazonas and 34.3% in Parà, representing an increase of 1.6 million tons or 46.8% in relation to the previous harvest [...]".

Still on the expansionist trend in sugarcane, Oliveira (2007, p. 2) points out:

> Faced with the current political and economic situation favorable to the ethanol market, this issue has become evident and has led to a strong movement by the sugarcane agribusiness on different fronts of expansion in the national territory, giving rise to new geographical and spatial (re)configurations, both in terms of capital and labor, with the state as its great ally.

On this issue, Oliveira (2009, p. 92) states:

> Since the 2000s, the territorial mobility of capital towards various states in the Center-South, such as the states of Minas Gerais, especially the Triângulo Mineiro (with a strong presence of mill owners from the Northeast), São Paulo, Goiàs, Mato Grosso do Sul, Mato Grosso, the north/northwest of Paranà and the north of Espírito Santo has not only been significant, but has also intensified. This movement has profoundly altered the geography of the countryside.

In this sense, Cordeiro (2008) states that, in this process of the expansion of sugarcane capital, there has been a notable trend towards the replacement of livestock areas with sugarcane areas in the Center-South of the country and the displacement of this activity to the Amazonian states. According to the data presented by the author, the states of Goiàs, Minas Gerais, Mato Grosso do Sul and Paranà had their pasture areas reduced by 20%, 19%, 16% and 14%, respectively, while cattle ranching in the Amazon saw a significant increase in headcount and pasture areas, comparing data from the IBGE agricultural census of 1995/96 and 2006. According to IBGE data, between 1995/96 and 2006 there was an increase in cattle farming in the states of the Northern Region, from 6.7 million to 17.5 million head, especially in the state of Parà.

In relation to the advance of sugarcane cultivation to the north of the country and the territorial dispute with the cultivation of other crops, Goes and Marra (2009) believe that the sugarcane plantations will not harm the Amazon rainforest, nor food production, considering that they occupy less than 2% of the country's agricultural area and that more than 99% of these plantations are located more than two thousand kilometers from the forest area.

Although there is controversy on this issue, the fact remains:

> The increase in food crop production between 1960-85 was lower than that of energy and export crops. While food crops increased by 138%, the increases seen in energy and export crops were 254% and 292% respectively (CARVALHO and CARRIJO, 2007, p. 8).

Magalhaes *et. al.* (1991) point out that the replacement of food crops in the last decades of the last century was not only by sugar cane, but also by soybeans, oranges and coffee, especially in the state of Sao Paulo.

Other data is also presented by the Cooperative of Sugar and Alcohol Producers (COPERSUCAR), disputing this claim, since only 1.3% of Brazil's agricultural area is occupied by sugar cane, while there is a large area of unproductive land occupied by livestock. He also points out that even in the state with the highest concentration of production (Sao Paulo), the area occupied by sugar cane is no more than 12% and sugar cane has replaced livestock rather than other crops.[6]

The area cultivated with sugar cane has grown in practically all the states in the Center-South region, with a significant increase in the 2000s in the state of Mato Grosso do Sul, concentrated mainly in the Ivinhema river basin.

Brazil is currently the world's largest producer of sugar cane, with the following states standing out: Paranâ, in the South; São Paulo, in the Southeast; Alagoas, in the Northeast; Goiâs and Mato Grosso do Sul, in the Midwest (SOUZA, 2014).

The spaces occupied by sugarcane plantations undergo spatial transformations rapidly and the interest in serving capital takes priority over the problems generated by this new production. The socio-spatial effects are felt to the extent that there is greater penetration of economic enterprises and large companies incorporate the productive space so that they can meet demand and, quickly, the sugarcane plantations gain ground on the old production. As a result, new public policies have been created to cater for this booming market.

> With the intensification of the use and exploitation of land and human labour in recent years, the new territorial and productive dynamics of sugarcane agribusiness in Brazil consolidate the understanding that capitalism is also manifesting its most advanced form in the countryside: industrial and financial capital, thus altering the dynamics historically built up in the areas where it is now homogenized and threatening different forms of land use (OLIVEIRA, 2009, p. 92).

Industrial capital is increasingly in charge of agricultural activities in the face of the economic interests of the national and international markets.

Economic growth is evident in the landscape through the geographic structures that represent progress, in this case the homogeneous landscapes of sugarcane production, constituting the

6 Information available at: http://www.copersucar.com.br. Accessed on: Mar. 2013.

main strategy used by the local government to attract investment and the construction of a discourse to be accepted and incorporated by agribusinesses and the local population (SOUZA, 2011, p. 117).

These new areas of sugarcane expansion are centers of attraction for workers from various regions of the country in search of better job opportunities. The sugarcane expansion fronts and the installation of agro-industries are attractions for migrants who often leave their families and move to work in the sugarcane fields or in companies in the sector, usually under precarious conditions.

From this perspective, Oliveira (2009, p. 65) points out:

> When Pro-Alcohol was implemented, as now, the government's justification for its support and, above all, for establishing policies that favored sugarcane agribusiness, was based on the "social reach" that the enterprise would have, generating jobs and income and promoting development by reducing regional inequalities. Without, however, taking into account that many of the jobs generated were temporary, as well as having an impact on the environment.

Thus, these regions are seen as "prosperous" and arouse the interest of large economic groups in terms of investment, as they receive government support for the installation of new branches in these areas and also attract workers, believing in the improvement of their living conditions by moving to these areas.

However, with the increasing use of machinery in sugar cane harvesting, this process has decreased in intensity. The mechanization of the production system exclusively serves the interests of capital, which aims to increase productivity and, consequently, profits. According to Oliveira (2009), mechanization reduces many jobs, considering that each harvester replaces, on average, the daily work of 80 men.

According to Costa (2008), investments, whether public or private, make it possible to gain agro-industrial productivity, increasing the average productivity of sugar cane and the average production of alcohol per hectare planted.

However, the modernization of the sugarcane industry did not happen quickly or evenly in the producing regions. The process is taking place in different ways and at different paces, and does not yet cover the whole process. In São Paulo, for example, sugar cane harvesting is mechanized on around 60% of properties, according to IBGE data.

On the advance of mechanization in the sugarcane sector, Oliveira (2009, p. 73) observes:

> [...] the mechanization of sugarcane cutting, which until a few years ago was a trend, is now a

concrete fact, even if it is still only partially adopted by the vast majority of sugarcane agro-industrial companies. In a few years' time, it could be the only option for harvesting sugarcane, both from an economic and ergonomic point of view (considering manual cutting), as well as from a legal and environmental point of view, since only mechanical cutting makes it possible to harvest without burning, even allowing the chaff to be used.

According to the author, although mechanization still suffers from some obstacles, such as the steepness of the terrain in some areas, these must be overcome, as the mechanized system is more economically advantageous. Cane harvesting without burning has an operating cost 32.74% lower than manual harvesting.

In the next chapter we will discuss how sugar cane entered the state of Mato Grosso do Sul and its socio-spatial effects on this part of the country.

CHAPTER III: CANE EXPANSION IN THE AGRICULTURAL CONTEXT OF THE STATE OF MATO GROSSO DO SUL

The main objective of this chapter is to characterize agricultural production in the state of Mato Grosso do Sul, showing how sugarcane plantations have expanded in this context, as well as their concentration in the Ivinhema river basin and current trends in the sector.

3.1. Occupation of the State of Mato Grosso do Sul

The territorial area that is now the state of Mato Grosso do Sul saw its occupation process intensify from the 1830s onwards, due to the displacement of people from within the Midwest region itself, mainly from Cuiabà and Santana do Paranaiba, who were interested in expanding the business related to cattle ranching (CAMPESTRINI and GUIMARAES, 1991).

Mining, which was initially responsible for the occupation of the Midwest, fell into decline in the 19th century and, as a result, the extraction of yerba mate and cattle ranching dominated the regional economic scene, initially without much expression, causing the region to experience a period of economic and demographic stagnation (ARBEX and OLIC, 1996).

The fact that really put this region at the center of discussion in the 19th century was the Paraguayan occupation and, consequently, the war (18641870). Towards the end of that century, the herb industry began to attract more attention, especially with the work of Companhia Mate Laranjeira, starting in 1872 (TEIXEIRA, 1989).

This activity was concentrated in the southeastern and southern portions of what is now Mato Grosso do Sul, occupying almost the entire Ivinhema river basin, and was an important factor in the settlement of the region (GUIMARAES and LEME, 1997). Companhia Mate Laranjeira dominated this activity almost exclusively in this area.

The company's operations in the region, exploiting the native herbs since the second half of the 19th century, had to do with the Platine market, as it was mainly targeting the Argentine market. The monopoly exercised by Companhia Mate Laranjeira worried the government, as it was believed that this activity did not contribute to the development of the national market and kept a lot of Paraguayan labor, making it difficult to bring in migrants from other regions of the country (MENEZES, 2012).

On this subject, Santos and Schlindwein (2011, p.6) state:

The exploitation of yerba mate in the south of Mato Grosso was linked to the Platine trade on the Paraguay River. In 1857 there were already exports from Corumbà to Buenos Aires and after the Paraguayan War exports intensified.

The conflicts that occurred with the proclamation of the Republic in 1889, between Republicans and Federalists, in Rio Grande do Sul, led to the revolution of 1892 to 1895, contributing to the migration of several southerners to the Midwest, settling mainly in the area of present-day Mato Grosso do Sul, occupying part of the Ivinhema river basin region, and dedicating themselves to cattle and herb activities. In addition to the southerners, the region also received migrants from São Paulo, Minas Gerais and the Northeast (ROSA, 1962).

Despite its effective occupation in the 19th century, the south of Mato Grosso, now Mato Grosso do Sul, only gained economic dynamism with the implementation of new transportation networks, linking it with important regions for the country's economy, such as São Paulo. In this context, the creation of the Noroeste do Brasil (NOB) railroad, now Novoeste, in the first decade of the 20th century was important, as was the opening up of the Paraguay river for navigation.

The opening up of navigation on the Paraguay River and the development of livestock farming strengthened the cities of Campo Grande and Corumbà, causing Cuiabà to lose its regional hegemony. Products such as standing cattle, yerba mate, rubber and other extractive products began to be exported, boosting the regional economy and bringing it closer to the country's major economic centers. The arrival of the NOB railroad further boosted cattle ranching, as cattle were now transported by rail (ROSA, 1962).

The development of cattle farming in the region attracted farmers from various parts of the country, mainly from Minas Gerais, São Paulo and Rio Grande do Sul. According to Rodrigues (1978), this encouraged the installation of charqueadas or saladeiros for the production of charque. With the implementation of the railroad, linking this region to the state of São Paulo, the cattle began to be sold in the São Paulo slaughterhouses and the charqueadas went into decline in the second decade of the 20th century.

Cattle ranchers from other regions of the country expanded their activities in the Cerrado of the Midwest, ensuring viable production in areas with low fertility soils, low use of technology, but vast natural pastures and abundant water. Along with cattle ranching, salt was also explored in the Pantanal (from the salt ponds), initially in the south of Mato Grosso, but without much success (RODRIGUES, 1978).

According to Teixeira and Hespanhol (2006), the biggest official incentive for settlement in the region was the so-called "march to the west" in the 1930s by the Getúlio Vargas government. This government made it possible to build roads and railroads, as well as eliminating customs barriers, facilitating trade between this region and other parts of the country, especially with the Southeast.

The practice of agriculture, however, was limited to a few areas with more fertile soils, particularly in the south of the state, especially in the municipality of Dourados.

In this context, the implementation of the National Agricultural Colony of Dourados (CAND) in 1943 was important. It brought about important demographic and economic changes, helping to fulfill the government's aim of regional integration (SANTOS, 2007).

The CAND played a fundamental role in attracting migrants from other Brazilian regions, who came in search of job opportunities and rural properties, forming a new colonization front, encompassing the areas of the current municipalities of Deodâpolis, Douradina, Dourados, Fàtima do Sul, Glòria de Dourados, Jatei and Vicentina.

> The Colony had an area of 267,000 ha which were divided into two zones separated by the Dourados River, the first located to the left of the river with 68,000 ha and the second to the right of the river with an area of 199,000 ha. [...] The area demarcated for the CAND was divided into 30 ha plots. For candidates to become settlers, the legislation required citizens over the age of 18 who were known to be poor and had agricultural qualifications. Although it was a nationalist project, CAND accepted foreigners with proven agricultural qualifications. Among them were the Japanese (MENEZES, 2012, p. 2-3).

The presence of the CAND contributed to a more equitable division of land in this part of the state, giving rise to small polyculture estates, unlike the rest of the state, which tended to develop monoculture estates.

In addition to CAND, the arrival of other colonizers was also important, such as Companhia de Viaçâo Sâo Paulo-Mato Grosso and Colonizadora Moura Andrade e Cia, whose activities attracted many people looking to buy land or provide services for these companies, resulting in the emergence of cities such as Nova Andradina and Batayporâ.

Initially, however, both livestock and crops were farmed in a rudimentary way, with cattle raised loose on natural pastures and crops planted without the use of modern techniques, using only animal-drawn plows and manual planters (TEIXEIRA, 2001).

3.1.1. The creation of the state of Mato Grosso do Sul

Discussions about the separation between the south and north of Mato Grosso began in 1896, when their main defense was related to the territorial extension of the state. However, this discussion became more intense during the Getúlio Vargas government, when there was a greater rapprochement between the people of southern Mato Grosso and those of São Paulo, especially during the 1932 revolution (RODRIGUES, 1978).

However, this only happened under Ernesto Geisel in the 1970s, when the state of Mato Grosso do Sul was officially created on October 11, 1977, by Complementary Law No. 31, and officially installed on January 1, 1979, as Santos and Schlindwein (2011, p. 7) show:

> During the military period, President Ernesto Geisel sent a bill to create a new state called Campo Grande, with the city of the same name as its capital. The law was approved by the National Congress and sanctioned by President Geisel on October 11, 1977, but the name of the state was changed to Mato Grosso do Sul, with Campo Grande remaining the state capital. On January 1, 1979, the government of Mato Grosso do Sul was officially installed, and Harry Amorim Costa was sworn in as governor.

The state borders Mato Grosso to the north; Goiàs and Minas Gerais to the northeast; Sao Paulo to the east; Paranà to the southeast (a total of 2,192.4 km of national border); as well as Paraguay to the south and southwest; and Bolivia to the west (a total of 1,520.3 km of international border).

The state currently has 79 municipalities spread over 357,145.532 km^2, covering approximately 22% of the Midwest region. Its absolute population, according to the 2010 IBGE Demographic Census, was 2,449,024 inhabitants, with 351,786 living in rural areas and 2,097,238 in urban areas. Its demographic density is 6.86 inhabitants/km2. In 2014, according to IBGE estimates, the state's population reached 2,619,657 inhabitants.

From the 1980s onwards, there was a great advance in the process of urbanization in the state, due to the intense rural-urban migration that took place during that period. Today, the urban population accounts for more than 85% of the total.

3.2. Agricultural production in Mato Grosso do Sul

The Cerrado's geoenvironmental conditions are favorable to intensive agriculture and livestock farming, according to Castro *et al.* (2010), but in Mato Grosso do Sul these activities have historically been extensive. Since the 1970s, with the modernization of agriculture coming to the region, these practices have been changing, but only timidly and limited to large properties.

Since the middle of the last century, the state of Mato Grosso do Sul has mainly developed grain agriculture, particularly soybeans, with a view to the foreign market. However, products such as cotton and beef were also important in incorporating the region into the national production system.

The Midwest region received important agricultural incentives in the 1970s through government programs to encourage the modernization of agriculture. Among the main programs were the Cerrado Development Program (POLOCENTRO), the Grande Dourados Development Program (PRODEGRAN) and the Pantanal Development Program (PRODEPAN). Other important incentives were instituted in the 1980s, through the Constitutional Financing Fund for the Midwest (FCO), created in the 1988 Federal Constitution (DUARTE, 1989).

Teixeira (1989) points out that these programs intensified the process of modernizing agriculture in Mato Grosso do Sul, integrating it into the national economy. Thus, despite the state maintaining the expansionist tendency of cattle farming, the production of cattle made significant progress from the 1970s onwards.

With the new investments in the region, there was a reordering of its productive space, with better use being made of the land, previously considered unproductive, to grow important crops. The new Brazilian agricultural frontier was thus defined during this period.

Carrijo and Miziara (2009, p. 96) state:

> [...] the state was an "inducer" of the transformation of the Midwest region, through targeted public policies. The creation of credit lines for rural producers, such as the National Rural Credit System, was essential for the application of technology in the countryside. Among other implications, we can see that this policy was responsible for the dissemination of technologies that enabled the transformation of the soil, promoting the occupation of the Cerrado [...].

In addition to the conversion of Cerrado soils into arable land, the rise in land prices in the South and Southeast in the 1970s led to an intense migration of people from these regions to the Midwest. Added to this were government incentives, mainly through the National Development Plans (PNDs), especially the II PND, which provided important lines of credit, including for land purchases, and resulted in other important regional development programs, mentioned in this chapter, which were highly subsidized and fundamental to the region's productive development (REZENDE, 2002).

The possibilities of expanding soya farming attracted producers from the south and southeast of the country, taking advantage of the incentives granted for soya production. In the 1980s, however, in the face of the recession the country went through, production of this product slowed down, but continued to expand (OLIVEIRA, 1993).

Thus, agriculture in Mato Grosso do Sul entered the process of modernizing agriculture in the 1970s and, even in the face of the crisis of the 1980s, the state made significant economic progress, with a significant increase in the production of grain and cattle, expanding the areas of crops and planted pastures. According to data from the Mato Grosso do Sul State Secretariat for the Environment, Planning, Science and Technology (SEMAC-MS), the state's economy grew by more than 8% a year in the 1980s.

The increase in the production of raw materials and the availability of official funds to finance private ventures have attracted many agro-industries to the state of Mato Grosso do Sul, especially grain crushers, meatpackers and distilleries.

According to the data contained in the IBGE's Municipal Agricultural Production (PAM), in the second half of the 1990s, the volume produced by the main crops expanded at an average annual rate of 3.44%. However, there were variations between types of production in terms of the rate of expansion during this period. Soy, for example, saw its production fall by 0.78% a year, despite having increased its area by 0.16% a year between 1995 and 1998. Corn, however, grew at an average rate of 11.6% per year during this period, despite also suffering a drop in production in 1997. The increase in its production area was 4.1% per year over the period.

At the turn of the 1990s and 2000s, the state's agricultural production began to expand again. According to SEMAC-MS data, between 1999 and 2003 alone there was a growth of 12.35%.

This growth, however, didn't last long and between 2004 and 2006 there was a reduction in production. This had a direct impact on agriculture's share of the state's GDP, which fell from 24.93% in 2003 to 14.52% in 2006 (SEMAC, 2011).

In the first decade of this century, important changes took place in the state's agriculture and livestock sector, with the advance of sugarcane cultivation and eucalyptus forestry, attracting large sugar and alcohol and pulp and paper companies.

At the same time, there was a decrease in the growth of cattle production and the number of cattle was reduced.

Soya, for example, accounted for 7.9% of national production in 2003 and 6.6% in 2004, according to SEMAC-MS data.

Sugarcane cultivation has expanded considerably in the central-south-eastern part of the state, mainly in the Ivinhema river basin, entering mainly pasture areas, but also areas of lavouras, and eucalyptus forestry has expanded in the central-eastern part of the state, replacing pasture areas (FARIA and FRATA, 2008; KUDLAVISK, 2011).

However, in this context, we can't ignore the lack of rainfall that occurred mainly between 2004 and 2006, which hampered crop production, as well as the outbreak of foot-and-mouth disease on the border with Paraguay in 2005, which affected the sale of beef produced in the state.

Economic progress in recent years has not altered the state's land structure, maintaining the high concentration of land ownership, with a predominance of monoculture estates and large tracts of unproductive land. Analyzing secondary data from the IBGE, from 1975 to 1996, we see a reduction in the number of establishments with an area of less than 100 hectares and an expansion in the proportion of area and quantity of establishments with more than 100 hectares.

In contrast to the period before 1990, between the last two IBGE agricultural censuses (1995/96 and 2006), there was a significant increase in the number of establishments of up to 100 ha, from 26,923 in 1996 to 42,673 in 2006, but no major change in the total area occupied, which went from 2.2% to 3.1% of the total.

The increase in the number of these smallholdings was due to the expansion of rural settlement areas in the state during this period. Data from the Mato Grosso do Sul Agricultural Development and Rural Extension Agency (AGRAER) show that more than twenty thousand families were settled in Mato Grosso do Sul during this period.[7]

Establishments with an area of between 100 and 1000 ha, however, did not undergo major changes and in the range of more than 1000 ha there was a reduction of more than 3% in their total and the area occupied by them fell from 78.4% to 76.9% of the total between 1995/96 and 2006.

Data from the 2006 IBGE Agricultural Census shows that there were 64,862 rural

7 Data obtained during a visit to AGRAER in the municipality of Anaurilândia-MS, in June 2013.

establishments in Mato Grosso do Sul, occupying 30,056,947 hectares. Of this total, 84.2% is occupied by pastures and 14.6% by crops.

Despite the increase in the number of establishments with an area of less than 100 hectares, the population and employment in the countryside continue to decline. According to SEMAC-MS data, the population employed in the primary sector in the state fell significantly in the 2000s. In 2005, this population represented 18.5% of the Economically Active Population (EAP), falling to 15.8% in 2009, while the secondary sector increased its share of the state's EAP, rising from 17.9% in 2005 to 20.9% in 2009. Even with the expansion of the sugarcane industry in the 2000s, employment in the countryside fell, signalling a trend towards increasingly mechanized production.

The amount of machinery and equipment in the countryside has increased significantly. According to data from the 2006 IBGE Agricultural Census, in that year Mato Grosso do Sul had 37,900 tractors, 4,279 harvesters and a fleet of 50,352 vehicles, made up as follows: 5,056 trucks, 10,896 utility vehicles, 9,009 cars, 7,420 trailers, 6,299 motorcycles, 64 planes and aircraft for agricultural use, 10,997 animal traction vehicles and 611 boats. Although we don't have the most up-to-date figures, SEMAC/MS guarantees that these numbers are more expressive today.

As for land use in the state of Mato Grosso do Sul, there have been some changes since 1975, as can be seen in Graph 3.

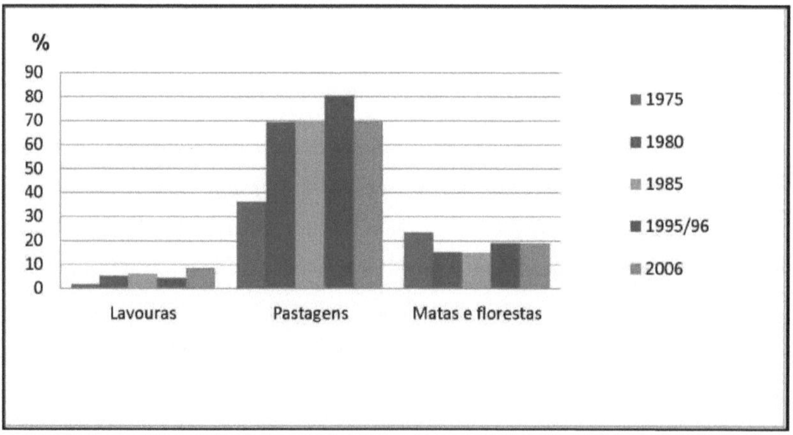

Graph 3: Mato Grosso do Sul: Land use between the 1970s and 2000s (%).

Source: IBGE - Agricultural Censuses of Mato Grosso do Sul for 1975, 1980, 1985, 1995/96 and 2006.

Organization: TEIXEIRA, J. C. - July 2014.

When we compare the data from the 1975, 1980, 1985, 1995/96 and 2006 Agricultural Censuses, we see a significant expansion in pasture areas between 1970 and 1995/96, representing 36.18% of the total in 1975, reaching 80.4% in 1995/96, but falling to 69.65% in 2006. This reduction represents 981,265 ha. Until the end of the 1990s, the state was in first place in terms of cattle farming, and with the expansion of sugar cane plantations and eucalyptus forestry in the first decade of the 21st century, pasture areas were replaced, which explains, in large part, the decrease in the relative importance of cattle farming in the state context.

These figures include both natural and planted pastures, but there has clearly been a reduction in natural pasture areas and an increase in planted pasture areas, especially with the introduction of brachiaria in the 1970s, a grass that is easily adapted to soils with low natural fertility (TEIXEIRA and HESPANHOL, 2006).

The decrease in pasture areas between the 1995/96 and 2006 Agricultural Censuses indicates the replacement of this production by new agricultural plantations, including sugar cane, as well as the monoculture of eucalyptus.

With regard to cattle numbers, despite the decrease in pasture areas, there was no reduction in total numbers between the last two censuses. The herd was 19,754,098 in 1995/96 and rose to 24,504,098 in 2006. However, in the last few years of the 2000s, according to data from the IBGE's Municipal Livestock Survey (PPM), the number of head of cattle fell and reached 22,354,077 in 2010. There is currently a downward trend in the herd, as many cattle farmers continue to lease their land for sugar cane plantations in the south and southeast of the state, and for eucalyptus cultivation in the east. Despite this, the state still has the 3rd largest cattle herd in the country, surpassed only by the states of Mato Grosso and Minas Gerais, according to IBGE's PPM data.

Despite the reduction in the number of cattle in the state, this activity is still very important in Mato Grosso do Sul's economy. It is mainly concentrated in medium-sized and large establishments, with more than 90% of production geared towards beef. Despite improvements in the breeding system, mainly since the 1990s, a large part of the herd is still raised in the extensive system.

As for the area under crops, there has been a considerable increase, although it is not very representative of the state as a whole. Over the last thirty years, there has been a steady increase until 2010, with the rate of growth remaining fairly stable until 2005. Between 2005 and 2010, however, crops grew in importance, reaching 9,683,742 tons of grain on 2,799,707 ha, according to PAM/IBGE data (Tables 3 and 4).

Table 3 shows that soybeans stand out, whose production was 1,322,082 tons in 1980 and reached 5,340,462 tons in 2010; corn, whose production was only 188,396 tons in 1980 and reached 3,782,946 tons in 2010; sorghum, whose production was 1,256 tons in 1980 and reached 162,153 tons in 2010. The biggest growth, however, was shown by sugar cane, whose production was 606,743 tons in 1980 and reached 34,795,664 tons in 2010.

Between 2000 and 2010 alone, sugarcane production grew by more than 25 million tons.

Among the products shown in table 3, wheat and rice saw their production drop significantly over this period. In 1980, the total amount of wheat and rice produced was 110,000 tons and 504,212 tons respectively, falling in 2010 to 74,201 tons and 142,668 tons.

Table 3: Mato Grosso do Sul: Production of the main crops between 1980 and 2010 (tons).

Crops	1980	1985	1990	1995	2000	2005	2010
Soy	1.322.082	2.558.720	2.038.614	2.283.546	2.486.120	3.718.514	5.340.462
Corn	188.396	327.334	595.718	1.435.151	1.069.571	1.291.901	3.782.946
Rice	504.212	323.993	182.458	239.269	226.649	224.831	142.668
Cotton	69.346	106.317	73.559	105.791	127.839	176.131	149.369
Wheat	110.000	317.644	204.035	19.786	34.712	136.410	74.207
Beans	23.507	29.882	33.966	23.590	10.019	23.595	31.937
Sorghum	1.256	16.522	5.241	5.136	83.527	178.715	162.153
Total grains	**2.218.799**	**3.680.412**	**3.133.591**	**4.112.269**	**4.038.437**	**5.750.097**	**9.683.742**
Sugar cane	606.743	3.170.806	4.193.288	4.922.386	5.837.456	9.513.818	34.795.664
Cassava	340.090	451.869	436.653	555.808	591.231	538.754	543.303

Source: SEMAC - Socioeconomic Diagnosis of Mato Grosso do Sul 2011, IBGE - Municipal Agricultural Survey 1980, 1985, 1990, 1995, 2000, 2005 and 2010.

Organization: TEIXEIRA, J. C. - June 2014.

Most of the state's agricultural production has always been in its southern portion, especially in the micro-regions of Dourados and Iguatemi. According to IBGE data, in 1980, this region was responsible for 56% of the state's grain production and, in 2010, for 64% of the total. The most important crops in this area during this period were soybeans, corn, manioc and sugar cane. However, sugar cane stands out, both in terms of territorial expansion and production volume.

Although cassava production appears as a traditional crop in various parts of the state, mainly in smallholdings, it has been expanding for commercial purposes. Production reached 543,303 tons in 2010, as shown in Table 3. The highest production is in the southeastern region of the state, especially in the municipalities of Ivinhema, with 104,160 tons and Eldorado, with production of 40,000 tons, according to PAM/IBGE data, mainly concentrated in establishments with less than 100 hectares.

Table 4: Mato Grosso do Sul: Harvested area of the main crops between 1980 and 2010 (ha).

Crops	1980	1985	1990	1995	2000	2005	2010
Soy	806.559	1.307.640	1.256.469	1.043.689	1.099.359	2.025.155	1.732.297
Corn	108.584	143.236	255.747	472.160	394.538	476.497	873.777
Rice	501.333	242.341	116.991	92.574	66.068	51.538	26.510
Cotton	44.615	66.619	44.570	60.011	48.450	63.718	38.740
Wheat	122.087	201.017	184.427	23.625	34.949	95.599	39.566
Beans	60.504	45.887	62.229	32.487	11.632	20.812	23.255
Sorghum	865	8.487	4.960	3.225	62.756	69.037	65.562
Grain harvested area	**1.644.547**	**2.015.227**	**1.925.393**	**1.727.771**	**1.717.752**	**2.802.356**	**2.799.707**
Sugar cane	11.671	50.650	67.358	74.815	98.938	136.803	399.408
Cassava	21.030	25.540	24.569	29.347	32.519	32.492	26.690

Source: SEMAC - Socioeconomic Diagnosis of Mato Grosso do Sul 2011, IBGE - Municipal Agricultural Survey 1980, 1985, 1990, 1995, 2000, 2005 and 2010.

Organization: TEIXEIRA, J. C. - June 2014.

From the data in Table 4, we can see a significant expansion in the areas cultivated with sugar cane in Mato Grosso do Sul over the period analyzed. In 1980, the area harvested with this product was 11,671 ha and, in 2010, it reached 399,408 ha, maintaining a higher growth rate than other crops.

The area harvested for soya in 1980 was 806,559 ha and in 2010 it reached 1,732,297 ha. The area harvested for this product has fluctuated over this period, with a small reduction between 2005 and 2010, but this has not affected the total produced, due to the increase in productivity.

Corn also saw its harvested area expand considerably during this period, from 108,584 ha in 1980 to 873,777 ha in 2010. Despite this great growth in the period analyzed, there were oscillations at the end of the 1990s, when this area shrank. However, in the 2000s this crop grew significantly again in the state, mainly due to the expansion of what is known as safrinha corn, which is grown outside its ideal period so as not to compete with soybeans.

Also noteworthy is the expansion of the area harvested for sorghum in the state, which in 1980 totaled only 865 ha and increased to 65,562 ha in 2010. This is due to the easy adaptation of this crop to Cerrado areas and the increased use of this product in animal feed in Mato Grosso do Sul.

Other important products, however, had their areas reduced during this period. Rice, for example, had a harvested area of 501,333 ha in 1980, down to 26,510 ha in 2010. Wheat also suffered a reduction in its harvested area, from 122,087 ha in 1980 to 39,566 ha in 2010. This was clearly reflected in the total produced, as shown above.

3.3. Sugarcane production in Mato Grosso do Sul

In the Cerrado areas, particularly in the states of Goiàs and Mato Grosso do Sul, sugar cane monoculture areas have been expanding since the end of the last century (CASTRO *et. al.*, 2010), but this process has intensified since the second half of the 2000s, as shown in the previous section. The increase in production is mainly due to the expansion of cultivated areas over pastures and other crops.

The state of Mato Grosso do Sul, despite the advance of soybean production since the 1970s, has expanded sugarcane cultivation by expanding this crop from the Southeast to the Midwest.

With regard to crop substitution, Nassar *et al.* (2008) point out that in the central region of Brazil, especially in the Cerrado areas, competition for space is mainly between soybeans and corn and between sugar cane and pastures. However, these authors state that sugar cane has also advanced into areas previously occupied by other crops, which we were able to verify during our fieldwork.

According to Castro *et. al.* (2010), the advance of sugarcane into the Midwest is part of the third cycle of expansion of this product in Brazil.

> The great sugarcane cycles can be summarized as threefold: the arrival and expansion of sugarcane in the Northeast since colonial times; Pro-Alcohol, which consolidated the Center-South; and now, the expansion of sugarcane to the north and northwest of the Center-South itself. The latter, which began in the early years of the 21st century, was favored, as before, by an international crisis linked to fuels, this time triggered by environmental demands for a new energy matrix aimed at controlling global warming (CASTRO *et.al.* 2010, p. 5).

Commercial sugarcane production began in Mato Grosso do Sul at the end of the 1970s, when the first distillery in the Central-West region was set up in the municipality of Pedro Gomes, but the sector was not very significant until the end of the 1990s (ASEVEDO and RIBEIRO, 2010). According to data from the National Balance of Sugarcane and Agroenergy, in 1984 the state produced alcohol, totaling 172,013m^3, but did not produce sugar, which began in 1989, totaling 21,925 tons.

According to Asevedo and Ribeiro (2010), in 1987, sugar cane production was in the municipalities of Pedro Gomes, Aparecida do Taboado, Nova Andradina, Brasilândia, Rio Brilhante, Sidrolândia, Maracajù and Navirai, with a planted area of 94,538 ha.

Until the end of the 1990s, however, there were no significant advances in production in the state, reflecting the crisis in the sector at a national level, when government funding became scarce (MONTEIRO, 2006).

In the 2000s, however, there was a strong expansion of sugarcane cultivation in Mato Grosso do Sul, especially in the second half of the decade, when it reached sixth place nationally among the states with the highest production. Sugarcane plantations advanced over pasture areas and, to a lesser extent, over crop areas, as we have seen.

This was due to major investments in the sugar-energy sector, including by important companies that have acquired land in the state or taken over vast tracts under lease and partnership. However, according to Azevedo (2010), production through partnerships is decreasing and production through leases is increasing.

The interviews we carried out in rural establishments located in the Ivinhema river basin show that there is a tendency for leasing to grow for sugar cane production and other crops, given the higher average yield compared to livestock farming.

Despite the greater concentration of the sugar-alcohol sector in the municipalities located in the Ivinhema river basin, Azevedo (2010, p. 58) points out:

> [...] after the initial boom in sugar cane expansion and the establishment of sugar and alcohol production units, there is a redirection of future projects to set up plants in less explored parts of the state, which, due to less favorable natural and economic conditions, have not yet been targeted by investors, such as some regions in the east and center-north, not to mention the emblematic Pantanal.

The growth of sugarcane cultivation in Mato Grosso do Sul raises questions about the influence of this expansion, with its exploitative model, on the concentration of land in the state and the increase in economic, social and environmental impacts.

The expansion of the sugar-energy sector is part of an extensive process of capitalist domination in the countryside, through agribusiness, which has taken hold in some parts of Brazil. In the case of Mato Grosso do Sul, agribusiness has excelled in recent years in sugarcane cultivation in the central-south and southeastern portions of the state and in eucalyptus forestry in its eastern portion, as mentioned above.

In both the sugarcane and eucalyptus sectors, large national and foreign economic groups are operating with huge resources in order to obtain ever greater production, extending their territorial dominance and earning significant income and profits in recent years. In the sugar-alcohol sector, the Odebrecht, Biosev and Adecoagro groups stand out, with the largest number of plants installed.

Thus, as Azevedo (2010) points out, the state offers the attractions desired by these groups, such as abundant land to support the expansion of their activities; attractive land prices, compared to other producing states; weak union organization; the state government's interest in hosting agro-industrial enterprises, exerting an attraction on companies.

Tax incentives are one of the main attractions for the installation of these companies and, given the interest on the part of the public authorities in hosting these agro-industries, what Santos and Silveira (2001) call the "fiscal war" takes place between the states and municipalities, explained by Pereira *et. al.* (2007, p. 10) as follows:

> Whenever a large enterprise wants to set up shop, one of the factors that can contribute to the decision on where to set up is the tax incentive packages and benefits granted by the local, state and federal authorities. In some cases, the race to attract companies to their territory leads to a kind of fiscal war, which occurs when states start offering increasingly attractive incentive packages in an attempt to bring companies to them. They do this by giving up revenue, making donations, among other things. The public interest in attracting companies is explained by the social and economic benefits they potentially represent, the main one being job creation.

In this sense, in Mato Grosso do Sul, the state and municipal governments offer attractions that are considered important for these companies, such as tax exemptions and land donations. In general, this has become a requirement of the companies themselves in order to set up in the state.

Leasing land for sugarcane production is becoming increasingly common in Mato Grosso do Sul, as the income from leasing is higher for this type of activity. According to Backes (2008), the rent for sugar cane production varies between R$400.00 and R$450.00 per hectare per year, while for soybean production it is around R$250.00

per hectare per year.

In our field research, we found that in the Ivinhema river basin region, the rent for sugar cane is on average R$400.00 per hectare per year. Although soybean production has become attractive in recent years, given the improvement in prices for this product on the market, leasing to the mills is still more advantageous when comparing the risks of production. However, some sugarcane producers turned to soybean farming in the 2014-15 harvest, a fact we'll discuss in Chapter 5.

Pereira *et. al.* (2007) state that the main factors encouraging the leasing of sugarcane plantations in Mato Grosso do Sul, in the

In the 2000s, there were crises in beef cattle and soybeans. In the mid-2000s, some beef exports were closed due to outbreaks of foot-and-mouth disease and the fall in the price of meat.

In addition, the repeated crop failures and the fall in soybean prices also led producers to opt for sugarcane. However, as we've said, the current reality is different, as there has been a resumption of investment in soybean farming, given the recent changes in the market for this product.

In relation to the price of land in Mato Grosso do Sul, Pereira *et. al.* (2007) made a comparison between the main producing municipalities (in Mato Grosso do Sul and São Paulo) and concluded that in the state of São Paulo land is up to three times more expensive than in Mato Grosso do Sul. This is yet another attractive factor for sugarcane producers and entrepreneurs in the sugar-alcohol sector to invest in the state.

In terms of soil and climate, the state has good conditions for the expansion of this crop, which is also important for the producer. The Cerrado's soil, despite its low natural fertility, is well drained and there are no major obstacles for the roots, as it is deep and stores water easily. The terrain, with few undulations, facilitates mechanized work, reducing production costs (MONTEIRO, 2010).

As for the infrastructure for transporting the product, the state still has problems, but due to its proximity to important consumer centers such as São Paulo, Minas Gerais and Paranà, this doesn't become such a significant barrier.

According to data from the Mato Grosso do Sul State Enterprise Management Agency (AGESUL), the road network in the state of Mato Grosso do Sul comprises 65,612.5 km of highways, of which 4,571.9 km (7.0%) are federal, 15,540.9 km (24.2%) state and 45,176.8 km (68.8%) municipal. Of this total, only 11.2% is paved, which corresponds to 7,338.8 km.

Law 328/1982 regulates the installation of sugar-alcohol industries in the state, protecting the Pantanal of Mato Grosso do Sul. Pereira *et al.* (2007, p. 18) present the main articles of this law:

Art. 1 - The installation of alcohol distilleries or sugar mills and similar is prohibited in the area of the Pantanal Sul-Mato-Grosso, corresponding to the area of the hydrographic basin of the Paraguay River and its tributaries, delimited in accordance with Annex I.

Art. 3 - The rights of the industries referred to in the previous articles which, on the date of publication of this law, are already up and running are guaranteed, subject to their operation being subject to compliance with the pollution control regulations in force.

The expansion of the installed capacity of the alcohol distilleries or sugar mills referred to in Article 1, which are already installed and operating on the date of publication of this Law, shall be prohibited.

The federal government's Ecological-Economic Zoning of Mato Grosso do Sul (ZEE-MS) and Sugar Cane Agro-Ecological Zoning (ZAE) public policies delimit the areas for the advance of sugar cane plantations for use by distilleries. Thus, in principle, the area of the Paraguay river basin is protected, i.e. the Pantanal portion of the state.

With job creation as its main argument, the state government tried to change the law in 2005, but with little success, as Pereira *et. al.* (2007, p. 20) show:

In 2005, the then governor of the state, José Orcirio Miranda (Zeca do PT), tried to expand the area allowed for the installation of sugar and alcohol plants. Through political lobbying and the mobilization of public opinion, the governor tried to change the law. However, in addition to pressure from certain Non-Governmental Organizations (NGOs), the State Assembly rejected this extension, based on CONAMA resolution 001/1985. It wasn't until the end of 2006 that this discussion was put back on the agenda, but this time it was changed to only allow the expansion of plants located in the Upper Paraguay river basin.

According to the Mato Grosso do Sul Bioenergy Association (BIOSUL), the sugar and alcohol companies based in the state are: Adecoagro - Vale do Ivinhema (Angélica), Adecoagro - Vale do Ivinhema (Ivinhema), Alcoolvale S/A (Aparecida do Taboado), Biosev Maracaju (Maracaju), Biosev Passa Tempo (Rio Brilhante), Biosev Rio Brilhante (Rio Brilhante), Bunge Monteverde (Ponta Porã), Companhia Brasileira de Açùcar e Alcool - CBAA (Brasilândia), Companhia Brasileira de Açùcar e Alcool - CBAA (Sidrolândia), DCOIL (Iguatemi), Energética Santa Helena (Nova Andradina),

Energética Vicentina (Vicentina), Fàtima do Sul Agroenergética (Fâtima do Sul), IACO Agricola (Chapadao do Sul), Odebrecht Agroindustrial - Costa Rica (Costa Rica), Odebrecht Agroindustrial - Eldorado (Rio Brilhante), Odebrecht Agroindustrial - Santa Luzia I (Nova Alvorada do Sul), Raizen Caarapó (Caarapó), Sao Fernando Açùcar e Alcool (Dourados), Tonon Bioenergia (Maracaju), Usina Aurora Açùcar e Alcool (Anaurilândia), Usina Laguna (Bataypora), Usina Sonora (Sonora) and Usinavi S/A (Navirai) (Map 1).

Of the 24 mills installed in the state, 17 are located in the Ivinhema river basin and, recently, three of these mills have ceased operating, limiting themselves to planting and supplying raw materials to other mills: CBAA, in Sidrolândia; CBAA, in Brasilândia; and, finally, Usina Aurora, in Anaurilândia.

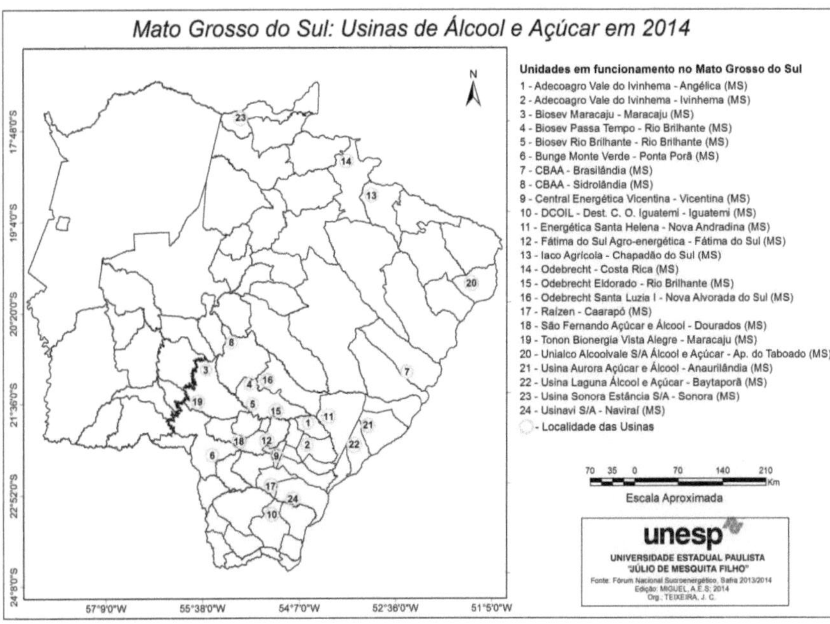

Map 1: Mato Grosso do Sul: Alcohol and sugar mills in 2014.

Oliveira (2009) states that the increasing presence of sugarcane agro-industries in the state and the expansion of sugarcane plantations in the 2000s signaled a trend towards a significant change in the productive arrangement, moving from cattle farming on planted pastures to sugarcane dominance, as can be seen in the area of the Ivinhema river basin.

Thus, the author points out:

> [...] the territorialization of the sugar-energy business in the state of Mato Grosso do Sul, as well as on the other fronts of recent expansion, reinforces the process of the internationalization of capital, expressed in the concentration and strengthening of groups through the incorporation of foreign capital via the acquisition of companies and/or shareholdings (OLIVEIRA, 2009, p. 205).

There is a common interest between the federal, state and municipal governments in strengthening sugarcane agribusiness, which leads us to think of a guaranteed future for this sector in the state, given the facilities granted by the government, even in the current crisis.

During a visit to the State Secretariat for the Environment, Planning, Science and Technology (SEMAC-MS), this was very evident, as they showed great expectations for the promising future of the sugarcane sector in the state and pointed out only positive aspects of this process. They say that the generation of jobs and income for workers and foreign currency for the state will improve people's quality of life. When asked about the possible environmental impacts generated, they say that there are no studies to prove such impacts and that they will not be any greater than those already generated by sojiculture and the extensive areas of pasture.

On the same visit, we were told that research is being carried out to make better use of sugar cane production in the state, with plants that are more resistant to pests and easily adaptable to any type of climate and its variations. Thus, the intention is to reach a yield of 200 tons/ha. Currently, the average yield is 87 tons per hectare planted.

Oliveira (2009, p. 212) says:

> A model that prioritizes exports, that incorporates producers into forms of business management, that diversifies products to gain competitiveness and guarantee new markets. A model that aims for productivity and profit; which, while claiming to be modern, incorporates new technological and environmental benchmarks and links up with the global capital market, from a social and labor point of view continues to exercise violence, subordination, overexploitation and practices similar forms of slave labor. Thus, even if it generates jobs and income, in the final analysis, the implementation of this development is much more favorable to capital.

In view of this growth in the sector, Mato Grosso do Sul reached 5th place in the *ranking of* the states with the highest sugarcane production in the country in 2010, when it produced around 35 million tons and reached an average productivity of 87 T/ha.

IBGE data shows that this production is concentrated in three mesoregions: Dourados, Iguatemi and Nova Andradina. In 2010, the production of the three mesoregions together amounted to 77.6% of the state total, with Dourados standing out with around 52%, or more than 18 million tons.

The average growth in sugar cane production in Mato Grosso do Sul over the last ten years has been higher than the national average. In the 2000s, the state's sugar cane production grew by 496.08%, while Brazil's growth was 120%. Alcohol production grew by 486.5% in the state and 162.46% in Brazil, while sugar grew by 473.45% and 109.51% in Brazil (SEMAC, 2011).

Sugarcane production in Mato Grosso do Sul, despite undergoing some variations, continues to expand and is concentrated mainly in the central-southern and southeastern portions of the state, i.e. in our study area, the Ivinhema river basin, which we will characterize in the next chapter.

CHAPTER IV: CHARACTERIZATION OF THE IVINHEMA RIVER HYDROGRAPHIC BASIN IN THE STATE OF MATO GROSSO DO SUL

We began this chapter by discussing the importance of studies focusing on river basins and their evolution in the field of Geography, placing this study in this context. However, our main focus is the characterization of the study area, the Ivinhema river basin, within the scope of the research, briefly addressing its physical aspects, based on the bibliographical survey of the area, and presenting the profile of the landowners and the characteristics of the rural establishments, according to the fieldwork.

4.1. Watersheds as a subject of study

The delimitation of a space for research is common practice in geographic science and is usually one of the first concerns of researchers, because empirical investigation is an important part of most research.

Watersheds are used as a cut-off point for research in the field of Geography, especially Physical Geography. In these areas, the interrelationship of physical elements favors studies of these spaces.

However, the territorial cut-off is also used in socio-spatial studies, which mainly seek to address the economic aspects of this space. Anthropic action is incorporated as another element, but the difficulties in carrying out research for Human Geography in these territorial portions of space lie in the justification for choosing these areas, considering that it is a natural cut-off, not so common for socio-economic research, which more often uses political-administrative cut-outs, such as states, municipalities or those delimited by the IBGE for statistical purposes, especially geographic micro- and meso-regions.

Throughout its history, and even today, Geography has been divided into Physical and Human, and this dichotomy is criticized in academic circles. The emphasis given to research ends up maintaining this separation and the delimitation of the space for the development of research follows this trend. From this perspective, the hydrographic basin is the most common area of research in the field of Physical Geography.

However, social facts play a major role in the organization of space and, even when the delimitation of the research takes into account only natural elements, human organization is essential for understanding geographical processes and phenomena.

In this sense, Santos (1982) states that spatial formations are, in fact, socio-spatial formations and that, based on human actions, the second nature or humanized nature is formed. In social organization, there is always economic interest in natural elements and nature is subject to the dictates of society.

From this perspective, Lima (2005, p. 177) points out that:

> In short, it can be said that while Physical Geography works with nature as a "system", Human Geography sees it as an "object". In other words, for Human Geography there is a socio-spatial system in which nature is just another element and for Physical Geography, on the other hand, there is a natural system in which society is an element.

Using a natural system for a socio-spatial approach is no simple task. The aim is to study the organization of space in a hydrographic basin from a human and economic perspective, even though this approach often goes beyond its physical limits.

However, studies of anthropogenic action in river basins have become more recurrent, mainly broadening discussions on the different ways of managing these areas, especially the formation of Basin Committees.

> Brazil's 1988 Federal Constitution established the hydrographic basin as the basis for regional planning. Law 9433/97 created the National Water Resources Management System (SNGRH), which consists of various decision-making levels at national, state and local (basin) level. At this last level, the Hydrographic Basin Committee was created, which is a decentralized and participatory body for discussion and deliberation, made up of representatives from different sectors of society that have an interest in the basin. The implementation of this management system is still relatively recent, but from it we can see that the problems related to water resources have led to changes in institutions and social practices and consequently have an influence on the organization of the geographical space represented by the basin (LIMA, 2005, p. 179).

Since the formation of watershed committees, studies on these areas have become increasingly important, as their elements must be studied taking into account their interactions and the proper management of these areas depends on knowledge of the physical and socio-spatial processes in these environments.

Thus, as Lima (2005, p. 180) demonstrates:

> This opens up a new dimension for Human Geography studies, with the hydrographic basin as a justifiable cut-off point. The watershed is born out of a naturalistic, geosystemic approach, but takes on a new meaning when society individualizes it according to its interests and defines a system of actions and created objects in this space that must be in balance with the natural

system.

Despite the importance of human geography studies focusing on river basins, problems arise when these studies need to analyze secondary data. This data is collected and published on the basis of municipal boundaries and aggregated into micro-regions, meso-regions and States of the Federation, sections that do not normally coincide with the boundaries of the river basins and, in themselves, also have very fluid boundaries. We ran into this problem in our research, because ten of the twenty-five municipalities studied are not entirely within the Ivinhema river basin.

Therefore, when we present the secondary data related to the research, we consider that they refer to the municipalities that are part of the basin and include part of the information from municipal areas outside the selected hydrographic basin.

4.2. Geographical aspects of the Ivinhema river basin - MS

The Ivinhema river basin is one of the nine sub-basins of the Paranà river basin in Mato Grosso do Sul and stretches from the center to the southeast of the state. It lies between latitude 20° 51' and 23° 14' South and longitude 52° 21' and 55° 57' West. It has a territorial area of 46,414.553 km^2 or around 4.64 million hectares and borders Paraguay to the west and the states of São Paulo and Paranà to the east (Map 2).

Map 2: Mato Grosso do Sul: Ivinhema river basin

The Ivinhema River (Figure 2), the main river in this basin, rises from the junction of the Vacaria and Brilhante rivers and runs for 205 km until it flows into the Paranà River.

This basin covers the municipalities of Anaurilândia, Angélica, Bataypora, Deodàpolis, Douradina, Dourados, Fàtima do Sul, Glòria de Dourados, Itapora, Ivinhema, Jatei, Novo Horizonte do Sul, Rio Brilhante, Taquarussu and Vicentina. Those with part of their areas within the basin are: Antonio Joao (41.5%), Caarapó (62.5%), Juti (40%), Laguna Carapa (33.9%) Maracaju (74.7%), Navirai (64.2%), Nova Alvorada do Sul (59.0%), Nova Andradina (78.5%), Ponta Pora (74.0%) and Sidrolândia (63.8%). Of the total areas of the twenty-five municipalities, 79.4% are within the Ivinhema basin, totaling 46,414.553 km^2.

Figure 2: Ivinhema River in the municipality of Ivinhema - MS.

Source: Fieldwork (Photo by the author - June 2013).

According to the 2010 IBGE Demographic Census, the population of the municipalities that make up the Ivinhema River basin totaled 654,704 inhabitants (26.7% of the state's population). The three municipalities with the largest populations are: Dourados (196,035 inhabitants), Ponta Pora (77,872 inhabitants) and Navirai (46,424 inhabitants). In terms of land area, the municipalities of Ponta Pora (5,330.448 km^2), Maracaju (5,299.184 km^2) and Sidrolândia (5,286.405 km2) stand out. The municipality with the smallest territorial area in the basin (33.9%) is Laguna Carapa (Table 5).

Table 5: Municipalities located in the Ivinhema river basin - MS: population, area and percentage of the area in the basin.

Municipalities	Population*	Total area (km^2)**	Area within the basin***	Area in the basin (%)
Anaurilândia	8.493	3.395,437	3.395,437	100,0
Angelica	9.185	1.273,268	1.273,268	100,0
Antonio Joào	8.208	1.145,175	475,340	41,5
Bataypora	10.936	1.828,024	1.828,024	100,0
Caarapó	25.767	2.089,600	1.307,820	62,5
Deodapolis	12.139	831,211	831,211	100,0
Douradina	5.364	280,787	280,787	100,0
Dourados	196.035	4.086,235	4.086,235	100,0
Fatima do Sul	19.035	315,160	315,160	100,0
Glòria de Dourados	9.927	491,748	491,748	100,0
Itaporà	20.865	1.321,814	1.321,814	100,0
Ivinhema	22.341	2.010,168	2.010,168	100,0
Jatei	4.011	1.927,946	1.927,946	100,0
Juti	5.900	1.584,540	634,050	40,0
Carapà Lagoon	6.491	1.734,068	589,030	33,9
Maracaju	37.405	5.299,184	3.961,130	74,7
Navirai	46.424	3.193,541	2.053,250	64,2
Nova Alvorada do Sul	16.432	4.019,323	2.372,940	59,0
New Andradina	45.585	4.776,002	3.750,240	78,5
Novo Horizonte do Sul	4.940	849,094	849,094	100,0
Ponta Porà	77.872	5.330,448	3.945,440	74,0
Rio Brilhante	30.663	3.987,397	3.987,397	100,0
Sidrolândia	42.132	5.286,405	3.375,740	63,8
Taquarussu	3.518	1.041,121	1.041,121	100,0
Vicentina	5.901	310,163	310,163	100,0
Total	**654.704**	**58.407,859**	**46.414,553**	**79,4**

Source: * IBGE - 2010 Demographic Census.

** IBGE - cities - In: http://www.ibge.com.br/cidadesat/xtras/home.php.

***OLIVEIRA, URCHEI and FIETZ (2000, p. 12).

Organization: TEIXEIRA, J. C. - June 2013

The geological structure of the area is part of the Paranà sedimentary basin and, according to the Water Resources Plan for the Ivinhema River Basin (2013), has the following geological formations on the surface: Alluvial Deposits, Caiuà Formation and Serra Geral Formation (Table 1).

Chart 1: Geological units of the Ivinhema river basin - MS.

Geological units	Characterization	Catchment area in the basin (%)
Alluvial deposits	They are mainly distributed on flood plains and along drainage channels. These deposits are predominantly composed of quartz sand, sand, gravel, silt, clay and, occasionally, peat.	6,39
Caiuà Formation	Formed in the Cretaceous, it has quartzose to subarcose reddish sandstones, bimodal (very fine and coarse), with angular to subrounded grains in the fine fraction and rounded grains in the coarse fraction, with variable sphericity. Selection is good in the fine fraction and moderate to good in the medium to coarse fractions. The predominant sedimentary structure is small to large tangential cross-stratification and, secondarily, lamination planoparallel.	43,90
Serra Geral Formation	Inserted in the Sâo Bento Group, established in the Cretaceous 138-129 Ma, by the Ar-Ar (Argon-Argon) dating method, it is characterized by basalts, black to dark grey in colour, thin to aphanitic, massive or with tonsils filled with calcite, quartz or clay minerals and with a fractured structure.	49,71

Source: Water Resources Plan for the Ivinhema River Basin (2013)

Organization: TEIXEIRA, J. C. - July 2015

As for the relief of this territorial area, according to Faria and Frata (2008), the plateaus on the western edge of the Parana River basin predominate in the south and the Maracaju and Dourados plateaus stand out in the north. The altitudes of this area vary from 240 to 600 meters.

The soil types that occur in the basin area are: red argisols, planosols, cambisols, purple latosols, red latosols, organosols, gleissols, quartz sands, as well as alluvial deposits (Map 3). Table 2 shows EMBRAPA's definition of each of these soil types.

Latosols are predominant and, according to Oliveira, Urchei and Fietz (2000), cover 84.7% of the basin, with purple latosols occupying 46.2% (21,466.76 km^2) and dark red latosols 38.5% of the area (17,883.89 km$^{2)}$. The former has greater natural fertility and is found in the highest part of the basin, while the latter stands out in the lower part of the basin and, despite being deep, requires some corrections to make good use of for crops. The other types of soil occur in much smaller areas and have no direct influence on crop areas.

Table 2: Soils of the Ivinhema river basin: definition and agricultural suitability.

Types of soil	Definition	Agricultural aptitude

Red clay soils	Soils made up of mineral material, whose distinguishing characteristic is the presence of a B textural horizon of low or high activity clay combined with low alkaline base saturation. Red Argisols have higher levels of iron oxides present in the original material in well-drained environments. The clay content in the subsurface horizon (red in color) is much higher than in the surface horizon.	Argisols with higher natural fertility (eutrophic), good physical conditions and more gentle terrain have greater potential for agricultural use.
Planosols	Imperfectly or poorly drained mineral soils, with an eluvial surface or subsurface horizon, with a lighter texture, which contrasts abruptly with the B horizon or with an abrupt transition combined with a marked difference in texture from the A to the immediately underlying, densified B horizon, generally with a high concentration of clay and slow permeability.	The potential for agricultural use of these soils is related to the environment in which they occur, especially the flat and gently undulating terrain. The limitations or restrictions are related to their slow or very slow permeability, and they are usually densified due to the accumulation of clay on their surface.
Cambisols	Soils made up of mineral material, which have an A or historic horizon with insufficient thickness to define the Organosols class, followed by an incipient B horizon.	In flatter areas, Cambisols, especially those with higher natural fertility, low activity clay and greater depth, have potential for agricultural use.
Purple latosols	Soils consisting of mineral material, with a brassolic B horizon immediately below any of the surface diagnostic horizon types, except historic.	Due to their good physical conditions, they have a high potential for agricultural use.
Red latosols	They are bright red in color, due to the higher levels and nature of the iron oxides present in the material originating in well-drained environments.	Because they are deep and porous or very porous, they present suitable conditions for good root development at depth, especially if they are eutrophic (high fertility). However, the nutritional potential of soils will be greatly reduced if they are alkaline, as there is the "chemical barrier" of aluminum which prevents root development at depth. If the soil is acidic, there is also a "chemical barrier", but in this case it is more related to the low sum of bases (especially calcium) than to aluminum saturation, which is not high in acidic soils. In addition to these aspects, they are soils which, under natural conditions, have low levels of phosphorus.
Organosols	Poorly evolved soils, with a preponderance of characteristics due to organic material, black or very dark gray or burnished, resulting from the accumulation of plant remains, in varying degrees of decomposition, in conditions of restricted drainage.	They present limitations or even restrictions to agricultural use, associated with the presence of high levels of sulphidic materials, salts and sulphur responsible for toxicity to most crops. Medium to high base saturation soils (eutrophic) indicate higher natural fertility, which increases the potential for agricultural

		use of these soils.
Gleissols	Soils with a glei horizon within the first 150 cm of the surface, immediately below the A and E horizons, or the H horizon (historical), with insufficient thickness to define the Organosols class.	It has low natural fertility (dystrophic) and may also have problems with acidity (very low pH) and high levels of aluminum, sodium (saline) and sulfur (thiomorphic). In terms of physical characteristics, they are poorly or very poorly drained soils under natural conditions.
Quartz sands	Soils originating from sandy deposits, with a sandy or sandy loam texture over a depth of at least 2m. These soils are essentially made up of quartz grains and are therefore practically devoid of primary minerals that are not very abundant.	Quartz sands are considered to be soils with low agricultural suitability. Continued use of annual crops can quickly lead to their degradation. Management practices that maintain or increase organic matter levels can reduce this problem.
	resistant to weathering.	
Alluvial deposits	They are most on flood plains, with muds (silt and clays) over large areas, and in mountain foothills as alluvial fans, with common deposits of fanglomerates and associated sands that reach a good areal expression and great thickness.	Alluvial deposits, which are usually very fertile for agriculture, especially those located on the banks of watercourses.

Source: EMBRAPA - Organization: TEIXEIRA, J. C - March 2015.

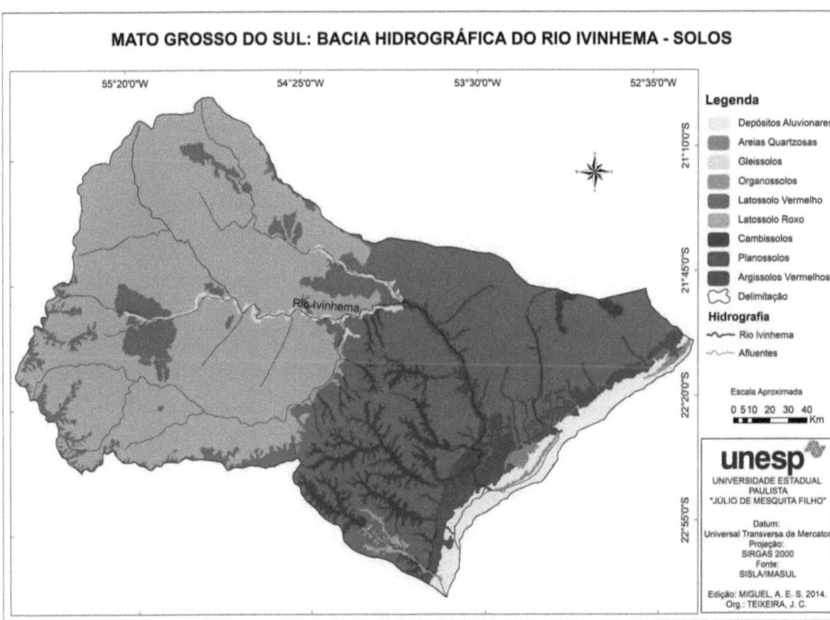

Map 3: Mato Grosso do Sul: Ivinhema River Basin - Soils.

In terms of climate, the Aw type in the north and northeast of the basin and the Cwa type in the rest of the basin stand out, according to the Koppen classification. The former is a humid climate with a dry winter and rainy summers and higher average temperatures; the latter, although also rainy with a dry winter, has lower average temperatures than the former. Average annual rainfall ranges from 1,400 to 1,700 mm (OLIVEIRA, URCHEI AND FIETZ, 2000).

As for the region's original vegetation, the Cerrados predominate (open wooded savannah in the highest part and dense wooded savannah in the lower altitude areas) and the semi-deciduous seasonal forest, to a lesser extent, in the southern part of the basin (OLIVEIRA, URCHEI and FIETZ, 2000). There is currently little left of this vegetation due to deforestation for pastures and crops.

4.3. Productive structure of the countryside in the Ivinhema River basin - MS and the profile of rural producers

The land ownership structure in the area where the Ivinhema river basin is located is not very different from that prevailing throughout the state of Mato Grosso do Sul, with a high concentration of larger establishments. However, the level of land concentration is lower than the state average.

Table 6 shows the number of rural establishments by size of area in the municipalities of the Ivinhema river basin, based on the 2006 IBGE Agricultural Census.

The data shows that 20.7% of the establishments have areas of less than 10 ha; 41.2%, between 10 and less than 100 ha; 28.8%, between 100 and less than 1,000 ha; 9%, between 1,000 and 10,000 ha and 0.3% of 10,000 hectares or more. Although there are more small and medium-sized establishments, the area of large establishments is larger, characterizing the poor distribution of land in the region and the dominance of the latifundia.

Some municipalities have a predominance of small and medium-sized farms. The municipalities of Caarapó and Vicentina have more than 40% of their rural establishments in the up to 10 ha stratum.

In the stratum between 10 and 100 ha, the municipalities of Glòria de Dourados, Novo Horizonte do Sul and Deodàpolis stand out; in the stratum between 100 and 1000 ha, there is a greater concentration in the municipalities of Maracaju, Rio Brilhante and Ponta Pora and between 1000 and 10,000 ha (9% of establishments), the municipality of Maracaju stands out, where these establishments represent more than 25% of the total.

Table 6: Municipalities in the Ivinhema river basin (MS): Number of rural establishments by size of area.

Municipalities	Less than 10 ha		10 to less than 100 ha		100 to less than 1,000 ha		1,000 less than 10,000 ha		More than 10,000 ha	
	No. of establishments	%	No. of establishments	%	No. of establishments	%	No. of establishments	%	No. of establishments	%
Anaurilândia	66	11,83	133	23,84	270	48,39	89	15,95	0	--
Angelica	178	32,13	238	42,96	108	19,49	30	5,42	0	--
Antonio Joâo	26	13,90	67	35,83	66	35,29	28	14,97	0	--
Batayporâ	201	35,25	196	34,39	133	23,33	39	6,84	1	0,18
Caarapó	424	41,37	318	31,02	235	22,93	48	4,68	0	--
Deodàpolis	185	21,89	518	61,30	141	16,69	11	0,12	0	--
Douradina	158	32,78	268	55,60	55	11,41	1	0,21	0	--
Dourados	818	37,13	829	37,63	479	21,74	77	3,50	0	--
Fàtima do Sul	298	37,67	454	57,40	39	4,93	0	--	0	--
Glòria de Dourados	141	17,56	566	70,49	96	11,96	0	--	0	--
Itaporâ	266	28,09	474	50,05	186	19,64	21	2,22	0	--
Ivinhema	442	31,87	759	54,72	162	11,68	22	1,59	2	0,14
Jatei	31	6,75	256	55,77	112	24,40	59	12,85	1	0,22
Juti	8	5,13	53	33,97	69	44,23	25	16,03	1	0,64
Laguna Carapâ	36	12,68	94	33,10	124	43,66	29	10,21	1	0,35
Maracaju	26	4,54	71	12,39	326	56,89	148	25,83	2	0.35
Navirai	89	18,58	174	36,33	138	28,81	75	15,66	7	1,12
Nova Alvorada do Sul	1	0,34	25	8,59	196	67.35	66	22,68	3	1,03
New Andradina	58	5,45	573	53,80	351	32,96	77	7,23	6	0,56
Novo Horizonte do	332	35,06	590	62,3	17	1,80	6	0,63	2	0,21

Sul										
Ponta Porã	29	4,65	168	26,92	316	50,64	104	16,67	7	1,12
Rio Brilhante	23	4,85	111	23,42	247	52,11	89	18,78	4	0,84
Sidrolândia	32	4,81	194	29,17	319	47,97	118	17,74	2	0,30
Taquarussu	126	30,88	195	47,79	69	16,91	18	4,41	0	--
Vicentina	362	42,14	444	51,69	52	6,05	1	0,12	0	--
Total/Average	4.356	20,68	7.768	41,19	4.306	28,84	1.181	9,02	35	0,26

Source: IBGE - Agricultural Census of the State of Mato Grosso do Sul, 2006.

Organization: TEIXEIRA, J. C - July 2014

During the fieldwork, we surveyed 75 farms in the Ivinhema river basin area, in the state of Mato Grosso do Sul, 10 of which had up to 10 ha; 8, between 11 and 100 ha; 17, between 101 and 500 ha; 18, between 501 and 1,000 ha and 22, over 1,000 ha.

The establishments surveyed are located in almost all the municipalities in the river basin, except for Antonio Joao, where, according to secondary data, we did not identify any sugarcane plantations (Table 7). Despite being a small sample , considering the total of 17,646 establishments, the investigation provided a good overview of the productive structure of the basin area.

Table 7: Number of rural establishments surveyed by municipality in the Ivinhema river basin - MS.

Municipalities	**No. of establishments Surveyed**
Anaurilândia	3
Angelica	3
Batayporâ	3
Caarapó	2
Deodapolis	3
Douradina	2
Dourados	6
Fatima do Sul	2
Glòria de Dourados	2
Itaporâ	3
Ivinhema	6
Jatei	2
Jutı	1
Laguna Carapâ	2
Maracaju	7
Navirai	2
Nova Alvorada do Sul	3
New Andradina	6

Novo Horizonte do Sul	2
Ponta Porã	2
Rio Brilhante	6
Sidrolândia	3
Taquarussu	3
Vicentina	1
Total	**75**

Source: Fieldwork (2013-2014).

Organization: TEIXEIRA, J. C. - March 2015.

We applied 75 forms with questions about rural establishments, producers and production, totaling 17 items, containing objective and open-ended questions.

We collected data and information on agricultural production, the structure of establishments and the socio-economic profile of producers. The intention is not, of course, to go into this, but rather to characterize the rural producers in this area, as they are the main agents in the production process.

Graph 4 shows the results of the survey in relation to the level of education of the rural producers surveyed in the Ivinhema river basin. Of the 75 farmers interviewed, most (29.3%) had incomplete primary education, demonstrating the low level of education among them. Only 2.7% of these producers declared themselves illiterate; 20% had completed elementary school; 17.3% had incomplete secondary school; 16% had completed secondary school; only 2% had incomplete higher education; 10.6% had completed higher education; and 1.4% had postgraduate degrees.

Graph 4: Ivinhema River Basin (MS): Level of education of the rural producers surveyed (%).

Source: Fieldwork.

Organization: TEIXEIRA, J. C. - June 2014.

Although the number of illiterate people is small, in general there is a low level of education. We observed that the predominant level is incomplete primary education, i.e. many of the interviewees have not completed the ninth grade.

From the fieldwork, we identified producers with a higher level of education among the medium and large producers, while the small producers have less formal education.

The parents have shown an interest in keeping the children at school. Some adults also study at night, taking part in the Youth and Adult Education Program (EJA), but admit that it is not easy to reconcile life in the countryside with formal education. In all the municipalities, the local government provides school transport, but only for the morning and afternoon periods. Thus, we see the need to set up rural schools for children, young people and adults, because education in the countryside is precarious, making it necessary to create public policies to serve this population. There are a considerable number of people in these establishments showing an interest in studying if rural schools were set up. According to Ferreira and Brandao (2011, p. 4) "[...] everything has been innovating in the countryside, except for education, except as a result of pressure from organized social movements".

Some of the farms surveyed are run by the owners' children and, in these cases, we observed a greater concern to keep their employees better informed and up to date, especially with regard to technological modernization. This is evident in the words of Mr. J. S C, 32, who took over the management of the rural establishment of more than three thousand hectares, where soya and sugar cane are grown and cattle are raised.

> [...] we can no longer think of a field without modern machinery and equipment. That's why my employees need to study and keep up with technological developments, otherwise they'll lose their place here in the field. And that's not what we want.[8]

With regard to the origin of the interviewees, we found that the majority came from the state of São Paulo (24%), while 18.7% were from the state of Mato Grosso do Sul itself. There are also people from Minas Gerais (17.3%), Paraná (13.4%) and Mato Grosso (10.6%), with those from other states accounting for less than 10% (Graph 5).

8 Excerpt from an interview conducted in March 2014 with Mr. J. S. C, a rural producer in the municipality of Dourados - MS.

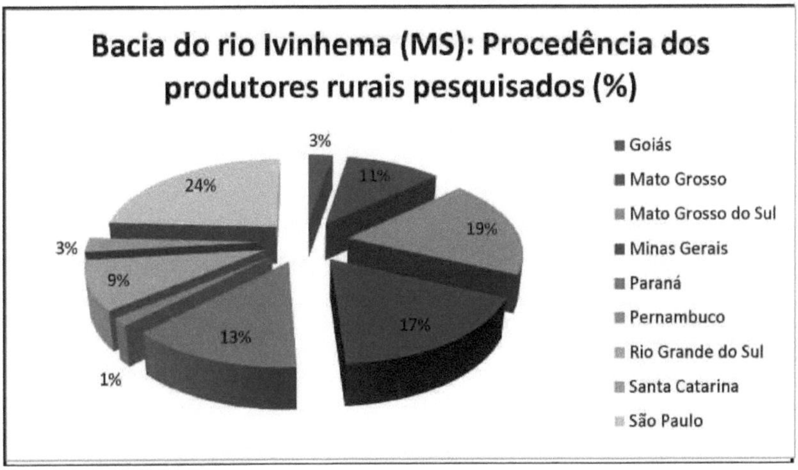

Graph 5: Ivinhema River Basin (MS): Origin of the farmers surveyed (%).

Source: Field research.

Organization: TEIXEIRA, J. C. - June 2014.

Most of the producers surveyed live in urban areas (69.3%), mainly in medium and large establishments, while 24% live on their own establishment and 6.7% in other places, such as other rural establishments.

It is common to find producers living in smallholdings, as production is usually family-based and requires their constant presence, confirming Castro's statement (2003, p. 23):

In family farming, the family usually lives on the farm and plays important roles in the production process, and working together with common goals differentiates them from production in the traditional capitalist system.

Before acquiring the property, 41.3% said they owned other properties, but 24% were rural employees, 17.3% were employed in the cities, 12% worked as tenants or partners and 5.4% were squatters.

Regarding the income of those interviewed, we identified: 12% had an income of more than 10 minimum wages; 46.6%, between 5 and 10 minimum wages; 24%, between 3 and 5 minimum wages; 16%, between 1 and 3 minimum wages; 1.4%, less than 1 minimum wage (Graph 6).

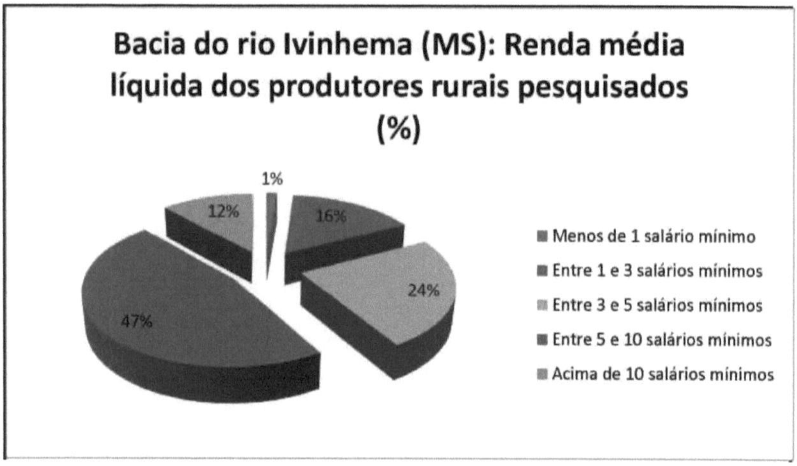

Graph 6: Ivinhema River Basin (MS): Average net income of the rural producers surveyed (%).

Source: Fieldwork.

Organization: TEIXEIRA, J. C.- June 2014.

The income of those surveyed does not only come from rural establishments, many have other sources. Only 5.3% of the producers declared that they had an income only from the production obtained on the farm; 24% also obtain income from other rural establishments; 21.3% are retired; 49.4% carry out other activities in the urban area.

It is common, for example, for small producers to work in sugar cane production areas, usually during the harvest period. While some members of the family work in other activities, some remain in the family farm looking after production, as stated by Mr. C. M. N, 62, a small producer in the municipality of Nova Andradina-MS.

[...] The farm can't be left without anyone, but it's also difficult to survive on it alone. So sometimes I stay and my children go to work in sugarcane, soybeans or corn and, at other times, I go, they stay and so life goes on [...].[9]

Table 8 shows the distribution of the resident population in the establishments surveyed, according to gender and age group, without any pre-established criteria. The number of people living in the establishments surveyed totals 660, 381 men and 279 women, with an average of 8.8 people per establishment.

We observed a predominance of people between the ages of 25 and 60, 50.8% of

9 Excerpt from an interview conducted in March 2014 with Mr. C. M. N, a small rural producer in the municipality of Nova Andradina - MS.

women and 43.8% of men, which is common considering that this is the age group in which people are most active.

In the age group up to 14 years old, 24.2% are male and 11.8% female, which shows that there are a lot of children in the countryside. This fact led us to ask about their permanence at school, but as we've already said, school transportation normally takes place in these municipalities, making it easier to keep children under 14 at school and also giving other age groups the opportunity to return to their studies.

Table 8: Ivinhema River Basin - MS: Population by gender living in the rural establishments surveyed.

N° OF PEOPLE	MEN		WOMEN	
	N°	%	N°	%
Up to 14 years old	92	24,2	33	11,8
Between 14 and 25 years old	68	17,8	55	19,8
Between 25 and 60 years old	167	43,8	142	50.8
Over 60	54	14,2	49	17,6
TOTAL	**381**	**100**	**279**	**100**

Source: Fieldwork.

Organization: TEIXEIRA, J. C. - June 2014.

In the establishments surveyed, 17.8% of men and 19.8% of women were aged between 14 and 25, and 14.2% of men and 17.6% of women were aged over 60.

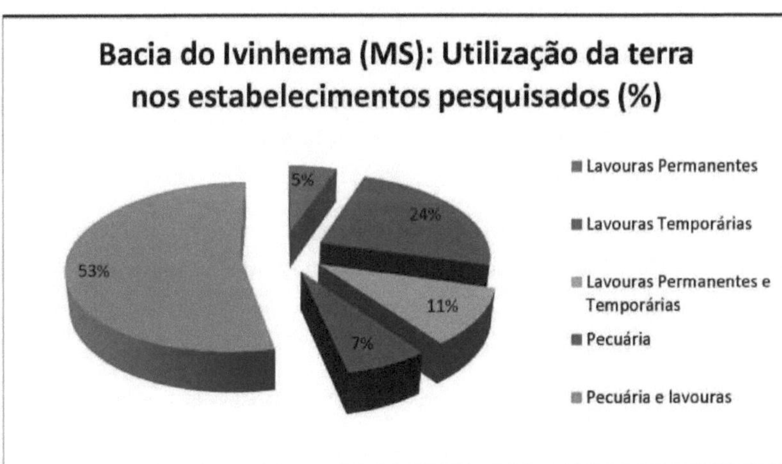

Graph 7: Ivinhema River Basin (MS): Land use in the establishments surveyed (%).

Source: Field research. Organization: TEIXEIRA, J. C. - June 2014.

As for the use of the land in the establishments surveyed, more than 50% is used for pasture for livestock and temporary crops, especially cattle, soybean, corn and sugar cane. Of the total, 24% are used only for temporary crops, in which there are a variety of products grown, but sugar cane predominates. Only 5.4% of the establishments surveyed cultivate only

products from permanent crops, 6.7% of them are used only for livestock and 10.7% have both permanent and temporary crops (Graph 7).

With regard to the workforce employed in the establishments surveyed, 57.3% have permanent employees, 9.3% have temporary employees, 20% have permanent and temporary workers and 13.4% use only family labor.

These figures only take into account the workers employed directly on the farm, and do not include outsourced workers, because on those farms where leasing takes place, there are also workers hired by the mills to work in sugarcane plantations, especially during the harvest period. In these establishments, we didn't find any children or indigenous people working at the time of the research, but we did get information from the producers themselves that indigenous people are hired, in small numbers, for the sugarcane harvest period.

As for the agricultural equipment used in the establishments surveyed, tractors stand out, present in 34 of the 75; 31 have a plow with mechanical traction, only five use a plow with animal traction and 14 have an irrigation system.

With regard to the use of fertilizers and pesticides, the figures are much higher, as we found these products in 92% of the establishments surveyed. There are also a considerable number of establishments that provide feed for livestock, especially during the dry season, totaling 73.4%.

Of the farms surveyed, 96% receive technical assistance, of which 22.7% is official, 57.3% is private, 9.3% is in-house and 6.7% comes from cooperatives. All the interviewees said that this assistance only takes place when requested by the owner, except in the case of areas leased to mills.

Most of the producers surveyed have not used any financing in the last 10 years (57.3%) and 42.7% have used these resources, mainly for planting temporary crops. Of all the producers, only 12% are members of cooperatives or producer associations, but 57.5% are members of trade unions.

As for the infrastructure of the establishments surveyed, 100% have electricity, 89.4% have piped water (from streams, artesian and semi-artesian wells), 28% have a telephone, 64% have warehouses for agricultural products, although not all of them are suitable for this purpose, and 17.4% have workshops for repairing agricultural machinery and implements.

With regard to housing conditions, we found that 84% of the houses are made of masonry, 10.7% are made of wood and 5.3% are mixed, i.e. partly made of wood and partly of masonry. Most of the houses are in a good state of repair (54.6%), while 33.4% are in a reasonable state of repair and 12% are in very poor condition. Most of the houses (81.4%) have between four and seven rooms.

Of all the homes, 41.4% have adequate sanitary facilities, while the rest have poor or incomplete facilities.

As for the durable consumer goods in the establishments, we found that 100% had a television, 61.4% a DVD player, 88% a fridge, 30.7% a freezer, 76% a stereo and 29.4% a computer.

The means of transportation used by residents of these rural establishments to travel to the cities are: bus (30.6%), car (29.4%), motorcycle (28%), cart (8%) and tractor (4%).

In general, we can conclude that there is a satisfactory standard of living in the establishments surveyed. Of course, this is a small sample considering the total number of farms in the Ivinhema river basin area, but according to Moraes (2011), the quality of life has improved in the Brazilian countryside in recent years, although it still needs more investment. In recent years, the countryside has ceased to be a place without infrastructure. Almost all properties have electricity (which allows the use of equipment and appliances), piped water and adequate sanitation facilities, but this still differs considerably between Brazilian regions and between rural establishments.

In the next chapter, we will analyze the advance of the sugar-alcohol sector in the Ivinhema river basin area in the context of regional agricultural production.

CHAPTER V: SUGARCANE EXPANSION IN THE CONTEXT OF AGRICULTURAL PRODUCTION IN THE IVINHEMA RIVER BASIN - MS

In this chapter we present an analysis of agricultural production in the municipalities of the Ivinhema river basin. However, our main objective was to verify the expansion of sugarcane plantations in this area from the 2000s onwards, when the sector expanded rapidly in this region. The analysis was based on the data we collected through fieldwork, when we visited rural establishments, sugar and alcohol mills, AGRAER, SEMAC-MS and INCRA, as well as secondary data obtained from the websites of the Brazilian Institute of Geography and Statistics (IBGE), the National Institute for Space Research (INPE), the Sugarcane Industry Union (UNICA), the Union of Bioenergy Producers (UDOP) and the Mato Grosso do Sul Agriculture and Livestock Federation (FAMASUL).

In this chapter, we also mapped the area studied, showing how land use has evolved in the basin and the evolution of sugarcane production in the period from 2003 to 2015, as well as the location of the sugar and alcohol plants in the area. Mapping land use has allowed us to visualize how sugarcane plantations have advanced over areas of other agricultural activities, especially pastures.

5.1. Agricultural production in the Ivinhema river basin - MS

The region in which the Ivinhema river basin is located is the most important agricultural producer in the state of Mato Grosso do Sul, with an emphasis on growing grains and sugar cane. Due to the presence of fertile soils, especially in the upper part of the basin where the purple latosol stands out, occupation was motivated by the expansion of agricultural activities.

According to Faria and Frata (2008), agricultural production in the basin area has always been sharply divided, because while the upper part is dominated by grain production, especially soya and maize, the lower part is dominated by cattle farming, with a predominance of pastures.

This region is currently experiencing a major expansion in sugar cane cultivation, associated with the arrival of several companies in the sugar-alcohol sector. This expansion began in 2003, but has intensified since 2008.

Permanent plantations are very limited in the basin area, and there are no large-scale crops. However, temporary crops occupy a large part of this space, especially soybeans, corn and sugar cane.[9]

Table 9 shows the harvested area of the main permanent crops in the municipalities of the Ivinhema river basin, as of 2004. These products are coffee, annatto and yerba mate. Among them, coffee stands out, although not very significantly.

Table 9: Municipalities in the Ivinhema river basin (MS): Area harvested from permanent crops between the years 2004 and 2012 (ha).

Permanent crops	2004	2006	2009	2012
Coffee (beans)	1.845	1.535	904	1020
Urucum (seed)	208	115	44	173
Yerba mate (green leaf)	147	124	66	61
Other products	51	45	90	92
Total	2.251	1.819	1.104	1.346

Source: IBGE - Municipal Agricultural Survey 2004, 2006, 2009 and 2012.

Organization: TEIXEIRA, J. C. - March 2015.

Analyzing the total areas harvested from permanent crops in the municipalities that make up the Ivinhema river basin, we saw a significant reduction, from 2,251 ha in 2004 to 1,346 ha in 2012. However, we noticed a difference in the expansion of production during the period, because while some crops showed fluctuations in the areas harvested, others showed a constant reduction.

[9] We are using the IBGE classification which classifies sugar cane as temporary, but we can consider it semi-perennial, as it allows up to 7 cuts per planting.

Coffee occupies the largest area of permanent crops in this region, but its area fell from 1,845 ha to 1,020 ha over this period. The total harvest of this product was 3,907 tons in 2004 and fell to 1,186 tons in 2012 (Table 10). Figure 3 shows a coffee plantation in the municipality of Ivinhema-MS.

Figure 3: Ivinhema - MS: Coffee plantations.

Source: Fieldwork (Photo by the author - January 2015).

The other permanent crops had negligible harvested areas throughout the period. Most of them are grown only for self-consumption in small rural establishments, as we saw during our fieldwork. In addition to the three products highlighted in the table, bay coconut, guava, orange, lime, passion fruit, heart of palm, rubber, tangerine and grapes are also grown, totaling only 92 ha in 2012.

Yerba mate, which was an important extractive activity and occupied a large part of this part of the state at the end of the 19th century and beginning of the 20th century, as we showed earlier, has lost its place. In 2004 the area harvested was 147 ha and in 2012 it was only 61 ha, with a constant reduction. Production in 2012 totaled 868 tons.

The areas once occupied by herbs have given way to soybean, corn and sugarcane plantations, according to Mr. W. S, a rural landowner in the municipality of Ponta Porâ, in an interview conducted in January 2015.

I stopped growing yerba mate and started growing soybeans and corn in lugar because the return is quicker. While corn and soybeans take six months to harvest, yerba mate takes two years. It's hard to wait all that if you have little incomo.[10]

Figure 4 shows a small urucum plantation in the municipality of Deodàpolis, another

10 Excerpt from an interview conducted in January 2015 with Mr. W. S, a rural producer in the municipality of Ponta Porà - MS.

permanent crop in the basin area.

Figure 4: Deodapolis - MS: Annatto plantation.

Source: Fieldwork (Photo by the author - September 2013).

The area harvested for annatto fell throughout the 2000s, but resumed its growth at the beginning of the 2010s. Total production of 125 tons in 2004 fell to 28 tons in 2009 and reached 145 tons in 2012.

According to information obtained from AGRAER in Ivinhema -MS, the municipality is the largest producer of annatto in the state and, although production is small and restricted to small rural establishments, there is a tendency for it to grow in the coming years, basically using family labor.

One product recently cultivated in the region is the rubber tree, currently found in the municipality of Nova Alvorada do Sul, but the total produced does not exceed 20 tons of coagulated latex, in an area of 10 ha. Other products, such as bay coconut and guava, have only appeared in recent harvests, according to IBGE data, but in very small areas, 14 ha and 40 ha respectively, destined for the free markets in the cities and to support their own families.

Table 10 shows the total production of permanent crops in the municipalities of the Ivinhema river basin. In this table, we have chosen not to group the products together, so that we can better verify the individual evolution of each production, even if in small quantities.

Through our fieldwork, we observed that the small areas of permanent crops in the Ivinhema river basin are located in smallholdings and that most of the produce is destined to support the family or to small businesses, mainly in the free markets of the cities.

Table 10: Municipalities in the Ivinhema river basin (MS): Total production of permanent crops between 2004 and 2012 (tons).

Permanent crops	2004	2006	2009	2012
Banana	75	182	575	158
Rubber (coagulated latex)				20
Coffee (beans)	3.907	2.218	587	1.186
Bay coconut*		26	80	80
Yerba mate (green leaf)	2.522	1.195	961	868
Guava			124	564
Orange	45			40
Lemon	15		10	10
Maracuja	67	4		69
Palm hearts		50		
Tangerine	15			
Urucum (seed)	125	19	28	145
Grapes	274	225	75	54

Source: IBGE - Municipal Agricultural Survey for 2004, 2006, 2009 and 2012. * in thousand fruits.

Organization: TEIXEIRA, J. C. - March 2014.

Temporary crops are the predominant crops in the Ivinhema basin. The three products with the largest areas harvested between 2004 and 2012, according to the publication Produçâo Agricola Municipal (PAM/IBGE), were soybeans, corn and sugar cane.

Soy is the agricultural product with the largest area of agricultural establishments in the municipalities of the Ivinhema basin, despite the fact that its harvested area fell slightly over this period, from 1,009,632 ha in 2004 to 993,507 ha in 2012. However, there was an increase in the volume produced, due to increased productivity. During this period, the total produced rose from 1,420,501 tons in 2004 to 2,733,160 tons in 2012, i.e. there was a reduction in the area cultivated, but a significant increase in productivity. Average productivity rose from 1,406 kilos/hectare in 2004 to 2,751 kilos/hectare in 2012.

Maize showed significant growth in its harvested area over the period analyzed, going from 423,194 ha in 2004, with a production volume of 1,481,648 tons, to 631,762 ha in 2012, when the total produced was 2,182,069 tons. It should be noted that this production refers to "off-season" corn, which is common in the basin area.

Sugarcane was the product with the greatest expansion in the area harvested in the basin in recent years. The data shows that in 2004 the area harvested for this product was 66,146 ha and in 2012 it reached 366,558 ha. The total sugarcane produced in 2004 was 5,472,893 tons and in 2012 it was 26,974,837 tons.

Figures 5, 6 and 7 show the three main crops grown in the Ivinhema river basin.

Figure 5: Maracaju - MS: Soybean crop.

Source: Fieldwork (Photo by the author - February 2015).

Figure 6: Taquarussu - MS: Corn crops. Source: Fieldwork (Photo by the author - September 2013).

Figure 7: Nova Andradina - MS: Sugarcane plantations.

Source: Fieldwork (Photo by the author - September 2013).

Table 11 shows the harvested area of the main temporary crop products (soybeans, corn and sugar cane) in the municipalities of the Ivinhema river basin. The areas of other products were grouped together because they occupy a smaller area, despite the variety of products. The sum of these areas is 80,817 ha, corresponding to less than 4% of the total.

Table 11: Municipalities in the Ivinhema river basin (MS): Area harvested from temporary lavouras between the years 2004 and 2012 (ha).

Temporary crops	**2004**	**2006**	**2009**	**2012**
Soybeans (in grain)	1.009.632	1.067.614	988.611	993.507
Corn (in grain)	423.194	419.844	531.087	631.762
Sugar cane	66.146	83.869	196.513	366.558
Other crops	103.049	145.882	87.302	80.817
Total	**1.602.021**	**1.717.209**	**1.803.513**	**2.072.644**

Source: IBGE - Municipal Agricultural Survey 2004, 2006, 2009 and 2012.

Organization: TEIXEIRA, J. C. - March 2015.

In Table 11, the products shown as "other crops" are: pineapple, grass cotton, peanuts, rice, oats, rye, beans, sunflower seeds, cassava, watermelon, sorghum, tomatoes and wheat. Among them, peanuts were the only one with an expanding area throughout the period analyzed, although not very significantly. Its area of 195 ha in 2004 increased to 816 ha in 2012, producing 1,151 tons of the product in shell.

During this period, some important products had their areas reduced, such as rice and beans. The area harvested for rice fell from 31,832 ha in 2004 to 16,520 ha in 2012. The area of beans harvested, from 17,867 ha in 2004, fell to 11,391 ha in 2012. The total amount of rice produced in 2004, 146,055 tons, fell to 91,282 tons in 2012. Bean production fell from 20,494 tons in 2004 to 14,737 tons in 2012.

According to information obtained from AGRAER in Dourados - MS, the lack of incentives and the discouragement of producers has reduced the area planted with rice by 75% over a period of 20 years in the state, and in less than five years, rice crops could disappear completely in Mato Grosso do Sul and, in the basin area, they should disappear in the next two years. The same is happening with bean crops because, according to the AGRAER technician, this product is very sensitive to high temperatures and farmers prefer to invest in soybean and corn production at the moment.

Cotton also saw its harvested area greatly reduced during this period. In 2004, it was 11,354 ha and in 2012, only 2,487 ha. Cotton production fell from 23,260 tons in 2004 to 8,785 tons in 2012.

There is a tendency for cotton to disappear from this part of the state, because the producers who are interested in this crop are in the northeast of the state, mainly in Chapadâo, São Gabriel do Oeste and Costa Rica.[11]

Sorghum's area fell from 11,519 hectares in 2004 to 1,776 hectares in 2012. Its production fell from 38,610 tons in 2004 to 4,948 tons in 2012. According to information obtained from AGRAER in Dourados - MS, this drop in production is associated with the reduction in livestock farming in the region, as sorghum is mainly used to feed cattle in the winter period, when there is a prolonged drought in the state.

Other temporary crops had their areas reduced during this period: sunflowers, watermelons, tomatoes and wheat. Some products fluctuated during the period, such as pineapple, oats and manioc, showing either growth or a reduction in the area harvested, but without much significance.

Table 12 shows the production volume of temporary crops in the municipalities of the Ivinhema river basin.

Table 12: Municipalities in the Ivinhema river basin (MS): Total production of temporary crops between 2004 and 2012 (tons).

Temporary crops	2004	2006	2009	2012
Pineapple*	455	878	1.569	1.368

11 Excerpt from an interview conducted in January 2015 with Mr. A. M. S, a rural producer in the municipality of Ponta Porâ - MS.

Herbaceous (seed) cotton	23.260	8.382	5.365	8.785
Groundnuts (in shell)	453	534	703	1.151
Rice (in husk)	146.055	107.258	105.865	91.282
Oats (in grain)	14.765	10.720	8.750	9.706
Sugar cane	5.472.893	7.229.794	18.095.327	26.974.837
Rye (in grain)	690			2
Beans (in bulk)	20.494	26.996	9.574	14.737
Sunflower (in bulk)		12.271	1.645	539
Castor bean (berry)	96	238		
Cassava	279.732	258.987	238.338	367.091
Watermelon	2.451	5.438	1.590	1.058
Melon		30		
Corn (in grain)	1.481.648	1.423.348	1.052.468	2.182.069
Soybeans (in grain)	1.420.501	2.424.500	2.131.988	2.733.160
Grain sorghum (in grain)	38.610	21.145	5.638	4.948
Tomatoes	2.700	1.599	1.504	1.175
Wheat (in grain)	118.441	37.383	54.191	29.520

Source: IBGE - Municipal Agricultural Survey for 2004, 2006, 2009 and 2012.

* in a thousand fruits.

Organization: TEIXEIRA, J. C. - June 2014.

During our fieldwork, we visited small establishments in the municipalities of Ivinhema and Deodàpolis that were investing in cassava production due to the guaranteed market for these products by the region's starch factories. In these municipalities, we observed a tendency for this crop to grow in the coming years, but the producers complain about the price paid by the mills.

[...] We manioc producers have to unite to demand a better price for our product. They need it and we have the product, but the price paid of 28 cents per gram of starch doesn't give us enough profit to grow our production, due to production costs. That's about 140 reais per ton, so it's impossible for us to continue in this business.[12]

According to information obtained from AGRAER in Ivinhema - MS, the municipality is the largest cassava producer in Mato Grosso do Sul and production in the Ivinhema hydrographic basin makes up around 60% of the total produced in the state.

Figure 8 shows a cassava plantation area in the municipality of Deodâpolis - MS.

12 Excerpt from an interview conducted in January 2015 with Mr. O. M. S, a cassava producer in the municipality of Deodapolis - MS.

Figure 8: Deodapolis - MS: Cassava farming.

Source: Fieldwork (Photo by the author - September 2013).

IBGE data reveals the disappearance of some crops in our study area during the period analyzed, such as castor beans, melons and triticale. Rye had its area drastically reduced, as in 2004 it had a harvested area of 690 ha and in 2012 only 2 ha were harvested.

Livestock farming is another prominent economic activity in the Ivinhema river basin area, but it is the activity that has lost the most ground in recent years.

Although cattle farming predominates throughout the state of Mato Grosso do Sul and in the Ivinhema river basin region, it has given way to crops, especially sugar cane. This explains the reduction in their numbers shown in Table 13.

According to the IBGE, the number of cattle was 4,713,349 in 2004 and fell to 3,800,967 in 2012 in the municipalities of this basin. Most of the herd is used for beef.

Figure 9 shows a rural beef cattle farm in the municipality of Bataypora - MS.

Figure 9: Bataypora - MS: Cattle farming.

Source: Fieldwork (Photo by the author - June 2013).

Table 13 shows the evolution of the main livestock numbers in the municipalities of the Ivinhema river basin between 2004 and 2012.

Table 13: Municipalities in the Ivinhema river basin (MS): Evolution of the main livestock numbers from 2004 to 2012.

FLOCKS	NO. OF HEADS			
	2004	**2006**	**2009**	**2012**
Asses	646	643	629	691
Cattle	4.713.349	4.394.247	3.991.490	3.800.967
Bubalinos	7.433	7.459	7.316	5.135
Goats	6.244	6.359	6.222	6.872
Horses	83.749	83.483	79.786	69.534
Muars	8.012	7.757	7.931	8.573
Sheep	86.404	93.222	100.310	123.191
Pigs	292.114	347.493	361.022	537.507
Total	**5.197.951**	**4.940.663**	**4.554.706**	**4.552.470**

Source: IBGE - Municipal Livestock Survey 2004, 2006, 2009 and 2012.

Organization: TEIXEIRA, J. C. - June 2013.

Milk production, which is more concentrated in small and medium-sized establishments, has also fallen in recent years. According to the IBGE, average production was 157,907 liters per day in 2004 and fell to 140,240 liters per day in 2012

(Graph 8).

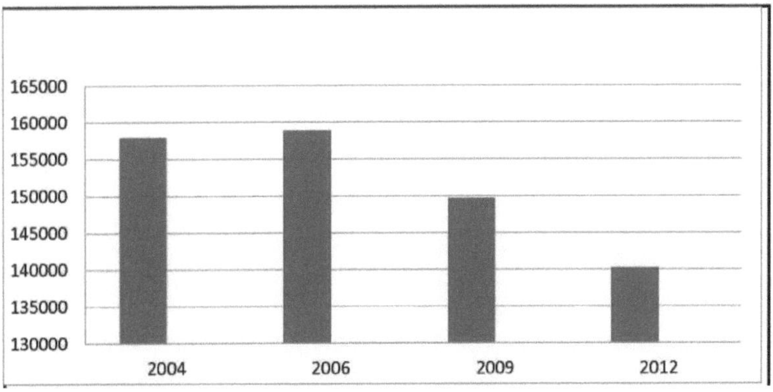

Graph 8: Municipalities in the Ivinhema river basin (MS): Average daily milk production (20042012).

Source: IBGE - Municipal Livestock Survey 2004, 2006, 2009 and 2012.

Organization: TEIXEIRA, J. C. - June 2014.

According to information obtained from AGRAER in Ivinhema, this municipality produces an average of 31,085 liters of milk a day, making it one of the largest producers in the Ivinhema basin area. Despite the growth of sugarcane production in the municipality in recent years, milk production has not been reduced because it occurs mainly in smallholdings, where sugarcane has not yet advanced.

In the basin area, the number of cattle has fallen and pastures are increasingly giving way to sugar cane, soybeans and corn, affecting milk production. This is the trend at state level, as we pointed out in the third chapter, even though there have been improvements in breeding techniques. Teixeira and Hespanhol (2014, p. 3) point this out:

> [...] new production techniques have been introduced in cattle ranching, but these advances are still limited, as many cattle ranchers in the Center-South of the country have leased their land for the cultivation of products that provide greater profitability per cultivated area, such as grains, notably soybeans and corn, sugar cane and forestry, mainly eucalyptus for the production of cellulose.

Graph 9 shows the reduction in pastures in the Ivinhema river basin area, based on the data generated from the mapping of land use and occupation, confirming the current trend in the sector.

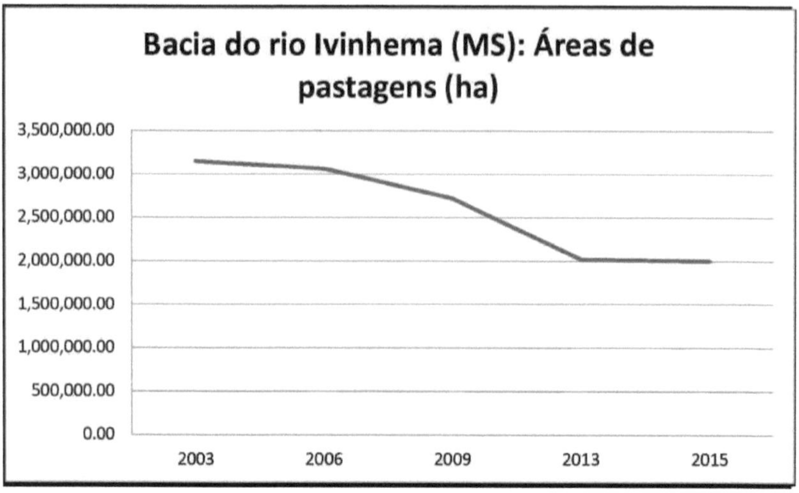

Graph 9: Ivinhema River Basin (MS): Pasture areas (ha).

Source: Data generated from mapping land use and occupation using satellite images.

Organization: TEIXEIRA, J. C. - June 2015.

The unfavorable scenario for livestock farming in recent years has led ranchers to abandon the activity in the state and migrate to other crops, causing the cattle herd to dwindle. According to cattle farmer J. A. M, from the municipality of Anaurilândia, the reason is the high cost of production and, consequently, the reduction in profits.

Raising cattle is becoming more and more expensive and laborious. I've done it all my life, since I was a boy... but I can see myself leasing the land for sugar cane or soy... sometimes I cry just thinking about it, but it's the reality and the trend in the Ivinhema valley.[13]

In this region of the state, the activity is not showing signs of recovery, even with the recent rise in the price of beef. According to information from SEMAC-MS, another issue that is detrimental to the increase in cattle numbers in the state is the growing slaughter of females, reducing the reproductive capacity of the herd.

According to the Federation of Agriculture and Livestock of Mato Grosso do Sul - FAMASUL[14] , the decrease in the number of slaughterhouses in the state reflects this trend. The number of slaughterhouses has fallen by 12% in the last three years. Today, the state has 22 slaughterhouses in operation. According to the agency, this is due to

13 Excerpt from an interview conducted in January 2015 with Mr. J. A. M, a cattle farmer in the municipality of Anaurilândia - MS.

14 Information available at: http:// www.famasul.com.br. Accessed on: Mar. 2015.

a reduction in the supply of cattle ready for slaughter in the state.

In addition to cattle, other herds have decreased: buffalo and horses. The former had 7,433 head in 2004 and fell to 5,135 in 2012; the latter had 83,749 head in 2004 and fell to 69,534 in 2012.

During the period, the only herds that grew were sheep (from 86,404 head to 123,191) and pigs (from 292,114 head to 537,507). Sheep have become a good income option in the state, due to the quick slaughter cycle, better use of the carcass, the improvement in the wool market and the presence of slaughterhouses in the state. With regard to pigs, the greater interest in this activity can be explained by the rise in the price of pork in recent years. However, the fieldwork did not provide us with more concrete answers to this question.

The herds of mules, goats and donkeys remained largely unchanged during the period analyzed.

In the rural establishments of the Ivinhema River basin, another highlight is the poultry sector.[15] In 2004, there were 14,921,370 birds and, in 2012, 15,634,809. This is due to the presence of poultry slaughterhouses in this area of the state, making it one of the areas with the highest production in Mato Grosso do Sul (MIZUSAKI, 2009).

5.2. Sugarcane production in the Ivinhema river basin - MS and its socio-spatial effects

In recent years, the state of Mato Grosso do Sul has seen rapid growth in sugarcane production, as we have seen above. This is especially true in the municipalities located in the Ivinhema river basin, which is the reason for this research. Of the 39 municipalities with sugarcane plantations in Mato Grosso do Sul, 24 are in this part of the state. Of the 25 municipalities that have all or part of their land in this basin, only Antonio Joâo does not grow sugar cane for commercial purposes.

However, this is a recent development. It was only in 2010 that we found sugar cane plantations in 24 of the 25 municipalities located in the area studied, according to data from the National Institute for Space Research (INPE).

As can be seen in Table 14, the area occupied by this crop has grown a lot in this basin. In the 2003-04 harvest, the area occupied by sugar cane was only 61,784 hectares, considering all the municipalities, and rose to 486,199 hectares in the 2012-

15 This includes hens, roosters, chickens, pullets and chicks (free-range and farm).

13 harvest.

The rate of growth was higher than the average for the main producing regions in the country and the area planted in these municipalities exceeds 74% of the state's total (FARIA and FRATA, 2008).

Table 14: Municipalities in the Ivinhema river basin (MS): Evolution of the area planted with sugarcane plantations between the 2003/04 and 2012/13 harvests (ha).

Municipalities	2003/04	2004/05	2005/06	2006/07	2007/08	2008/09	2009/10	2010/11	2011/12	2012/13
Anaurilândia						546	1.560	1.980	2.697	3.203
Angelica				1.496	3.641	12.239	18.760	23.285	30.116	36.120
Antônio Joâo										
Batayporâ						1.583	2.977	4.175	4.336	6.144
Caarapó						890	9.442	12.806	19.717	22.253
Deodàpolis					398	745	847	1.076	2.063	2.639
Douradina							157	157	157	157
Dourados				352	1.184	5.051	14.291	19.731	29.534	38.604
Fàtima do Sul						79	106	394	673	762
Glòria de Dourados						31	31	31	31	84
Itaporâ					2.880	4.420	6.245	6.308	7.735	7.936
Ivinhema					146	1.983	7.686	15.389	16.180	22.331
Jatei						788	2.202	2.253	2.673	3.722
Juti			720	781	780	976	2.053	5.773	9.206	11.834
Laguna Carapâ							3.279	9.861	10.712	11.344
Maracaju	11.553	11.553	16.307	17.574	19.389	25.092	33.245	35.398	36.760	37.316
Navirai			13.249	17.720	19.978	21.721	21.918	21.555	19.261	18.774
Nova Alvorada do Sul	12.450	12.451	13.418	14.964	19.403	22.332	28.450	37.004	53.295	71.966
New Andradina	16.683	16.777	18.559	17.881	20.341	22.593	21.743	23.959	26.957	31.601
Novo Horizonte do Sul								4.445	4.445	4.802
Ponta Porâ						1.346	8.222	13.687	21.505	30.985
Rio Brilhante	13.629	13.631	15.301	22.181	36.400	59.453	75.708	79.098	83.195	87.289
Sidrolândia	7.469	7.625	8.475	9.228	11.631	15.718	27.298	28.219	28.669	28.391
Taquarussu						144	3.232	3.421	3.670	4.437
Vicentina						1.618	1.710	1.910	2.893	3.505
Total	**61.784**	**62.037**	**86.029**	**102.177**	**136.171**	**199.348**	**291.162**	**352.278**	**416.480**	**486.199**
Total MS	**105.361**	**111.455**	**159.806**	**182.061**	**226.958**	**310.711**	**425.539**	**502.063**	**571.316**	**656.683**

Source: CANASAT - sugarcane mapping via observation satellite images - available at:

http://www.dsr.inpe.br/laf/canasat/. Organization: TEIXEIRA, J. C. - June 2013

The municipalities with the highest sugar cane production in the region, exceeding 30,000 hectares of planted area, are: Rio Brilhante (87,289 ha), Nova Alvorada do Sul (71,966 ha), Dourados (38,604 ha), Maracaju (37,316 ha), Angélica (36,120 ha), Nova Andradina (31,601 ha) and Ponta Porâ (30,985 ha).

The municipalities with the smallest areas cultivated with sugar cane, not exceeding four thousand hectares, are: Glória de Dourados (87 ha), Douradina (157 ha), Fâtima do Sul (762 ha), Deodâpolis (2,639 ha), Anaurilândia (3,203 ha), Vicentina (3,505 ha) and Jatei (3,722 ha).

Only five of these municipalities produced sugar cane in all ten of the agricultural years analyzed. They are: Maracaju, Nova Andradina, Nova Alvorada do Sul, Rio Brilhante and Sidrolândia. With the exception of Sidrolândia, they are all among the largest producers in the basin. This is justified by the fact that these municipalities received the first mills installed in this territory, which demanded greater sugarcane production.

The municipality of Rio Brilhante has two plants belonging to the Biosev group and one to the Odebrecht group, which demand a large amount of raw materials. As a result, this municipality has more than 80,000 hectares of sugarcane, corresponding to approximately 18% of the area planted with this crop in the Ivinhema river basin.

The municipality of Nova Alvorada do Sul is the second largest sugarcane plantation in the basin, meeting the demand for raw materials from the Odebrecht group, which controls the Santa Luzia I mill in operation in the municipality, as well as supplying raw materials to the group's other mills when necessary.

Although sugarcane production in the municipality of Dourados only began in 2006, it has expanded considerably, overtaking the municipality of Maracaju, a traditional producer, with more than 38,000 hectares planted. This advance was due to the increased interest of rural producers in leasing their land to mills in the region, given the installation of the Sao Fernando mill in this municipality and the proximity of the Bunge Monteverde and Raizen Caarapó mills.

The municipalities of Angélica and Ivinhema saw their areas planted with sugar cane expand rapidly with the arrival of the Adecoagro group's mills. In 2006, Angélica had only 1,496 hectares planted with sugar cane and in 2012 it had 36,120 hectares. In Ivinhema, where the plant was set up more recently, production began in 2007 with a small area of 146 ha, and in 2012 it had 22,331 ha of this crop.

Of the twenty-five municipalities in this basin, only Antonio Joao has no commercial sugarcane plantations and two of them (Douradina and Glória de Dourados) had negligible sugarcane production, with 157 ha and 84 ha planted, respectively, in 2012. Another municipality that stands out is Navirai, which, despite having the Usinav-SA mill on its territory, has seen a decrease in the area planted with sugar cane since 2009.

We can see from the INPE data that in the 2009-2010 harvest there were 21,918 hectares planted with this crop in Navirai and in the 2012-2013 harvest it fell to 18,774 hectares. In part, this can be explained by the advance of other crops in the municipality, such as soybeans and corn. According to the IBGE, the area harvested for corn in Navirai in the 2009 harvest was close to 34,000 hectares and expanded to more than 45,000 hectares in 2012. Soybeans grew a little less, from around 47,000 hectares harvested in 2009 to around 50,000 hectares in 2012. In addition, Usinav-SA has faced financial problems in recent years and has slowed down its raw material processing.

The municipalities of Anaurilândia and Caarapó only started producing sugarcane in 2009 with the arrival of the Raizen (Caarapó) and Aurora (Anaurilândia) mills. However, the growth of this crop in the municipality of Caarapó was greater than in the municipality of Anaurilândia. In 2009, Caarapó had 890 ha planted with sugar cane and, in 2012, it reached 22,253 ha. In Anaurilândia this growth was slower, going from 546 ha in 2009 to 3,203 ha in 2012. The justification given by the producers on this issue refers to the economic power of the two groups, with the Raizen group investing more in the sector.

In the municipalities of Bataypora and Ponta Pora, sugarcane plantations began to be planted in 2008, when the Laguna (Bataypora) and Bunge Monteverde (Ponta Pora) mills were set up. However, in the municipality of Ponta Pora, production expanded more quickly, considering that in 2008 sugar cane occupied 1,346 ha and in 2012, 30,985 ha. During our fieldwork, we found that part of this production is also supplied to the Sâo Fernando mill, located in the municipality of Dourados.

In the municipality of Bataypora, the planted area of 1,583 hectares of sugar cane in 2008 has grown to 6,144 hectares in the 2012-2013 harvest. The lack of raw material in the municipality means that the Laguna mill buys this product in the municipalities of Taquarussu and Nova Andradina, and there is even competition between it and Energética Santa Helena for the lease of areas for cultivation. According to the producers, the Laguna plant offers a monthly lease payment, attracting more landowners, as the payment made by the Santa Helena plant is quarterly.

Thus, we can see that sugar cane production in these municipalities was linked to the need to produce raw materials to meet the demand of the sugar and alcohol mills set up in the region. As these companies set up in the basin, sugarcane production grew rapidly, reinforcing Monteiro's (2010, p. 36) assertion:

> The sugar and alcohol mills have been expanding into the interior of the country, especially into the Center-West, causing a restructuring of the productive space in order to guarantee their raw material, sugar cane. Industrial capital is once again in charge of Brazilian agricultural production.

As we have already stated, of the 24 plants in operation in 2014 in the state of Mato Grosso do Sul, 17 are located in the Ivinhema river basin, as shown in map 4 and table 15.

Map 4: Mato Grosso do Sul: Ivinhema River Basin - Location of alcohol and sugar mills.

Table 15: Ivinhema river basin (MS): Sugar and alcohol plants in operation in 2014.

Production unit	Municipality	Start of operation	Economic group
Adecoagro - Ivinhema Valley	Ivinhema	2012	Adecoagro

Adecoagro - Ivinhema Valley	Angèlica	2008	Adecoagro
Aurora Sugar and Alcohol	Anaurilândia	2009	Aurora
Biosev Maracaju	Maracaju	1982	Biosev
Biosev Passa Tempo	Rio Brilhante	1982	Biosev
Biosev Rio Brilhante	Rio Brilhante	2008	Biosev
Bunge Monteverde	Ponta Porà	2009	Bunge
Brazilian Sugar and Alcohol Company - CBAA	Sidrolândia	1977	José Pessoa
Energètica Santa Helena	New Andradina	1978	St. Helena
Energètica Vicentina	Vicentina	2008	Toledo
Fatima do Sul Agroenergetics	Fatima do Sul	2011	Gadotti
Laguna Alcohol and Sugar	Batayporà	2009	Laguna
Odebrecht Agroindustrial - Eldorado	Rio Brilhante	2009	Odebrecht
Odebrecht Agroindustrial - Santa Luzia I	Nova Alvorada do Sul	2009	Odebrecht
Raizen Caarapó	Caarapó	2009	Raizen
Sào Fernando Sugar and Alcohol	Dourados	2009	Bertin
Tonon Bioenergy	Maracaju	2009	Tonon

Source: Mato Grosso do Sul State Secretariat for the Environment, Planning, Science and Technology - SEMAC-MS. Union of Bioenergy Producers - UDOP.

Organization: TEIXEIRA, J. C. - October 2014.

Table 15 shows that most of the mills in operation in 2014 in Mato Grosso do Sul began their activities in the state in the 2000s, mainly in 2008 and 2009. Only four of these mills operated in this region in the last century.

In the first decade of the current century, as the country's sugar-alcohol sector advanced, new economic groups from this sector began operating in Mato Grosso do Sul. Between 2008 and 2012 alone, the area of the Ivinhema river basin received 13 plants.

According to Monteiro (2014), the Environmental Licensing Manual of 2004 and the Ecological Economic Zoning of 2009 were important in guiding the installation of these companies. According to this author, these guidelines were fundamental to the growth of the sugarcane agro-industry in the state, defining the environmental rules and incentives for increasing activity.

Also important were the incentives from the state government and the financing acquired, mainly from the BNDES.

As a result, agricultural activity was brought closer to industrial activity, and production benefited in the state itself. This justifies the growth in sugarcane production in the Ivinhema basin during this period and, consequently, the increase in ethanol and sugar production, offered to the domestic and foreign markets.

In the 2000s, new economic groups with national, international and mixed capital began operating in Mato Grosso do Sul. There were important mergers and acquisitions for the sector in the state.

According to SEMAC-MS, there were forecasts of much greater growth in this area, as 24 new projects to set up plants in this area did not materialize.

The Adecoagro group has invested in two nearby towns (Angélica and Ivinhema), inaugurating the Vale do Ivinhema plant, with units in both towns. In Angélica, the company has been operating since 2008, with the capacity to produce four million tons of sugar cane in an area of approximately 50,000 hectares, with the aim of producing sugar and ethanol. In Ivinhema, the plant started up in 2012, initially aiming only to produce ethanol.[16]

The Raizen group emerged from the merger of Cosan and Shell in 2011. The group controls the Raizen Caarapó mill, in the municipality of Caarapó, but has more than twenty mills in Brazil. The Caarapó mill has the capacity to process 2.1 million tons of sugar cane.[17]

The Biosev group controls two plants in the municipality of Rio Brilhante (Biosev Rio Brilhante and Biosev Passa Tempo) and one in the municipality of Maracaju (Biosev Maracaju). The group was formed in 2009, following the merger of LDC Bioenergia, owned by the French group Louis Dreyfus Commodities, and Santelisa Vale. The Biosev Maracaju and Biosev Passa Tempo plants have been in the state since the 1980s, but were acquired by the French group LDC Bioenergia in 2007.[18]

The Odebrecht group controls three units in Mato Grosso do Sul, two of which are in the Ivinhema basin area: Odebrecht Agroindustrial Eldorado, in the municipality of Rio Brilhante, and Odebrecht Agroindustrial Santa Luzia I, in the municipality of Nova Alvorada do Sul. In the state, the group's processing capacity is 10 million tons of sugar cane.[19]

16 Information obtained during a visit to the Adecoagro plant (Angélica Unit) in September 2013, during fieldwork.
17 Information available at: http://www.raizen.com.br. Accessed on: June 2013.
18Information available at: http:// www.biosev.com. Accessed on: Jun. 2013.
19 Information available at: http:// www.odebrechtagroindustrial.com. Accessed on: Jun. 2013.

The Bunge group owns the Monte Verde mill, located in the municipality of Ponta Pora. This group is new to the sugar-alcohol sector. It entered the industry in 2006 and has made significant investments in the region. The Monte Verde mill is the second acquired by the group.[20]

The Tonon group owns a mill in the Ivinhema basin area, Tonon Bioenergia, located in the municipality of Maracaju. The group has an area planted with sugar cane of around 41,000 ha and has an installed processing capacity of 2.5 million tons. In 2013, the total sugar produced was around 170,000 tons and over 90,000 m^3 of ethanol.[21]

The Bertin group controls the Sao Fernando mill in the municipality of Dourados, which cultivates around 60,000 hectares of sugar cane. The mill has the capacity to process more than four million tons of sugar cane, producing approximately 300,000 tons of sugar and 150,000 m3 of ethanol.[22]

The Santa Helena group has only one mill, Santa Helena LTDA, which has been operating in the municipality of Nova Andradina since 1978. However, the group acquired the mill in the early 1990s. In 2013, it had its own planted area of approximately 17,000 hectares and produced around 240,000 liters of ethanol per day.[23]

The Laguna group is also new to the sugar-alcohol sector and has just one mill, Laguna Alcool e Açúcar LTDA, which began operating in 2009 in the municipality of Batayporâ. In 2013, this mill processed around 650,000 tons of sugar cane.[24]

The Toledo group owns a plant in the municipality of Vicentina, Central Energètica Vicentina, which began operating in 2008. It has a planted area of approximately 7,000 hectares and produces more than 500,000 tons of sugar cane.[25]

The Gadott group began operating in the municipality of Fàtima do Sul in 2011, setting up the Fàtima do Sul Agroenergética LTDA mill, with capacity to process around one million tons of sugar cane. It has a planted area of around 15,000 ha.[26]

The Aurora group recently entered the sugar-alcohol sector and set up the Aurora Açúcar e Alcool LTDA mill in the municipality of Anaurilândia in 2008. The mill had a

---

20 Information available at: http:// www.bunge.com.br. Accessed on: June 2013.
21Information available at: http:// www.tononbioenergia.com.br. Accessed on: June 2013.
22Information available at: http:// www.usinasaofernando.com.br. Accessed on: June 2013.
23 Information obtained from a visit to Usina Santa Helena in March 2013, during fieldwork.
24 Information obtained during a visit to Usina Laguna in March 2013, during fieldwork.
25 Information available at: http:// www.usinavicentina.com.br. Accessed on: June 2013.
26 Information available at: http:// www.udop.com.br. Accessed on: June 2013.

capacity to process 162,500 tons of sugar cane. In 2013, it produced more than 57,000 m^3 of ethanol and more than 80,000 tons of sugar.[27] However, in 2015, the plant ceased operating due to financial problems.

The José Pessoa group controls Companhia Brasileira de Açúcar e Alcool (CBAA), with two mills in Mato Grosso do Sul, one of which, CBAA-Sidrolândia, is in our study area, in the municipality of Sidrolândia. This plant had been facing financial problems in recent years and in 2014 it closed down completely.[28]

In the field research, we found that most of the farmers interviewed had leased their land to companies in recent years to plant sugar cane. Of all those interviewed, 56% leased in the last ten years, while 32% planted sugar cane on their own, usually with a contract to supply some mill in the region.

> In Mato Grosso do Sul, there is a tendency to lease land for the expansion of sugarcane agribusiness and eucalyptus forestry in the state, as this guarantees the landowner an income from the land, without him having to worry about the losses of agriculture and the disadvantages of livestock farming (MONTEIRO, 2012, p. 32).

Most of the producers interviewed used to raise cattle as the main activity on their farms (81.4%) before sugarcane cultivation, while 12% worked with other crops, mainly soybeans and corn. This data confirms the greater expansion of sugarcane plantations over pasture areas.

Even with the end of the sugarcane industry, a large part of these producers still continued with other production alternatives on the farm, such as soy, corn and cattle.[29]

Figure 10 shows sugar cane and corn crops on the same rural establishment in the municipality of Nova Andradina.

---

27 Information obtained during a visit to the Aurora mill in December 2014, during fieldwork.
28 Information obtained during a visit to the CBAA-Sidrolândia plant in December 2014 during fieldwork.
29 Normally, the same area as corn is occupied by soybeans. In the first half of the year, safrinha corn is planted, and in the second half, the same area is occupied by soybeans.

Figure 10: Nova Andradina - MS: Corn and sugarcane crops on the same rural establishment.

Source: Fieldwork (Photo by the author - June 2013).

Most of the producers interviewed said that sugarcane plantations had replaced degraded pasture areas on their farms. Only 5.3% of those interviewed said that they had switched from soybean farming to sugarcane farming. Of all the producers who opted for sugar cane as a new economic activity, 74.6% said that the main reason for this switch was the expectation of greater income, given the feasibility calculation before choosing the new product. However, 34.7% said they were not fully aware of the possible environmental impacts of sugarcane cultivation.

As for the sugarcane production system in this area, most of it is mechanized and, according to the producers interviewed, this was mainly due to the advance of the system of leasing land to the mills. Approximately 75% of the establishments surveyed produce sugar cane using mechanized harvesting. Around 15% use manual harvesting without burning, but around 10% still use manual cutting with burning. Figures 11, 12 and 13 illustrate this reality.

Figura 11: Nova Andradina - MS: manual harvesting of sugar cane without burning.

Source: fieldwork (Photo: Wagner Suzano de Freitas - June 2013).

Figura 12: Nova Andradina - MS: Burning in a sugarcane field.

Source: fieldwork (Photo: Wagner Suzano de Freitas - June 2013).

Figure 13: Nova Andradina - MS: Machine used to harvest sugar cane. Source: fieldwork (Photo: Wagner Suzano de Freitas - June 2013).

In Mato Grosso do Sul, the burning of sugar cane has its limitations. Since 2010, it has been banned in 18 municipalities, all belonging to the Ivinhema river basin. They are: Anaurilândia, Angélica, Bataypora, Caarapó, Deodapolis, Douradina, Dourados, Fatima do Sul, Glória de Dourados, Itapora, Ivinhema, Maracaju, Nova Alvorada do Sul, Nova Andradina, Novo Horizonte do Sul, Rio Brilhante, Taquarussu and Vicentina.[30]

This ban has greatly reduced this practice in the basin area, however, there is resistance from producers to comply with the law and burning still occurs in these municipalities, usually at dawn. According to the Mato Grosso do Sul Environment Institute (IMASUL), the state government itself has taken a stand in favor of producers

30 Information available at: http://www.imasul.ms.gov.br. Accessed on: June 2015.

in the sector, appealing the decision of the Federal Public Prosecutor's Office, but without success.

Aerial spraying of sugarcane plantations is also common. According to the interviews conducted at the mills, this takes place in accordance with the law, within the established limits. During fieldwork, we heard complaints from small farmers in establishments located on the margins of these fields, saying that their crops are affected by the poisons dispersed by the planes.

The planes pass overhead dropping poison. It ends up here on my property and my manioc gets it all. You can't even think about agro-ecological production with the plant nearby. They've been to the Public Prosecutor's Office, but so far nothing has been resolved.[31]

Figure 14 shows the dispersion of poison sprayed by plane in an area of sugar cane in the municipality of Nova Andradina.

Figure 14: Nova Andradina - MS: aerial spraying of sugarcane fields. Source: Field work (photo by the author - December 2014).

Because our study area has important indigenous reserves, another problem encountered is the constant conflicts over land ownership between Indians, landowners and mill owners. The indigenous people denounce the invasion of their lands by agribusiness, especially the sugarcane industry. In these conflicts, there have been deaths of indigenous people and the issue has been discussed by the legislative, executive and judicial authorities, but the expected progress has not yet been made.

31 Excerpt from an interview conducted in December 2014 with Mr. J. M. S, a small producer in the municipality of Nova Andradina - MS.

Recently, the Minister of Justice visited the state to discuss alternatives for resolving the problem.

There are academic works that denounce indigenous exploitation in the sugarcane fields, but in our fieldwork we didn't find any indigenous people working in the fields, which is justified above all by the high level of mechanization currently present in sugarcane harvesting, which has considerably reduced manual cutting.

There are also conflicts with the landless, who are fighting against the territorialization of agribusiness and the advance of sugarcane plantations over food areas, as well as seeking to occupy land with legal problems or that is unproductive. During our fieldwork, we found some encampments on the sides of highways, usually close to soybean and sugarcane plantations.

According to INCRA, there are 82 rural settlements in the Ivinhema basin area, with 12,756 settled families, covering a total area of 397,656 ha. This total corresponds to 44.5% of the families settled in the state.

Family farming is very important as a producer of products that are of no interest to large producers, including basic foodstuffs, including milk. However, it has been facing problems related to production, mainly due to a lack of financial resources. According to D'Alkmin Filho and Teixeira (2012), some of these producers have acquired resources from PRONAF and supply their products to PNAE, but there is a lack of information about these programs, which leads to low participation. These producers eventually receive technical assistance from AGRAER.

Table 16 shows the settlements present in the municipalities of the Ivinhema river basin in 2014.

Table 16: Settlements located in the municipalities of the Ivinhema river basin - MS in 2014.

	Municipalities	Settlement names	Area (ha)	Settled Families	Date of creation
1.	Anaurilândia	Saint Irene	2.471	75	27/03/2000
2.	Anaurilândia	Hope	4.074	99	12/07/2004
3.	Anaurilândia	Santa Ana	2.894	42	30/10/2008
4.	Anaurilândia	Barreiro	3.570	86	23/10/2006
5.	Angelica	Angelica	1.825	105	12/07/2005
6.	Bataypora	Mercedina	803	56	10/07/1996
7.	Bataypora	São Luiz	1.599	108	22/07/1994
8.	Bataypora	Saint John	856	55	11/03/1996
9.	Caarapó	N. S. do Carmo	1.193	39	06/11/2000
10.	Dourados	Lagoa Grande	4.070	157	06/11/1997
11.	Dourados	Amparo	1.126	67	22/12/1997
12.	Glòria de Dourados	Aimoré	440	39	27/12/2006

13.	Ivinhema	San Sebastian	2.967	96	23/12/1998
14.	Jatei	New Hope	2.820	83	28/08/1986
15.	Juti	New Panambizinho	1.681	47	16/11/2004
16.	Juti	Santa Clara II	1.243	78	16/11/2004
17.	Maracaju	Cantagalo	1.250	44	22/09/2000
18.	Maracaju	Valinhos	2.033	85	21/12/2001
19.	Maracaju	San Francisco	140	7	04/06/2003
20.	Maracaju	Santa Guilhermina	7.994	223	22/12/1997
21.	Navirai	Juncal	2.453	113	31/08/2001
22.	N. Alvorada do Sul	Ranildo da Silva	2.842	174	10/08/2005
23.	N. Alvorada do Sul	Saint Lucia	1.168	70	30/10/2007
24.	N. Alvorada do Sul	Volta Redonda CUT	1.562	116	30/10/2007
25.	N. Alvorada do Sul	Volta Redonda FAF	403	23	24/12/2007
26.	N. Alvorada do Sul	Drinking fountain	1.429	99	23/04/2004
27.	N. Alvorada do Sul	Pam	5.029	115	23/12/1998
28.	N. Alvorada do Sul	Success	968	80	30/10/2007
29.	N. Alvorada do Sul	New Dawn	3.000	86	12/05/1997
30.	New Andradina	Teijin	28.497	1.068	26/07/2002
31.	New Andradina	Saint John	3.905	178	23/04/2004
32.	New Andradina	Green House	29.859	461	22/12/1987
33.	New Andradina	Santa Olga	1.492	168	28/06/2004
34.	New Horizon	New Horizon	16.580	681	10/04/1986
35.	Ponta Porâ	Aba da Serra I	726	39	05/12/2007
36.	Ponta Porâ	Itamarati II - CUT	3.919	258	30/09/2004
37.	Ponta Porâ	Aba da Serra II	347	17	24/12/2007
38.	Ponta Porâ	Itamarati II FAF	2.215	148	30/09/2004
39.	Ponta Porâ	Itamarati - CUT	6.332	285	29/12/2000
40.	Ponta Porâ	Itamarati - AMFFI	3.585	147	29/12/2000
41.	Ponta Porâ	Itamarati II - FETAGRI	24.619	589	30/09/2004
42.	Ponta Porâ	Corona	1.095	58	03/06/1997
43.	Ponta Porâ	Itamarati FETAGRI	8.164	388	29/12/2000
44.	Ponta Porâ	Dorcelina Folador	8.011	269	27/11/2000
45.	Ponta Porâ	Itamarati II FAF	1.238	110	30/09/2004
46.	Ponta Porâ	Itamarati - MST	7.226	322	30/09/2004
47.	Ponta Porâ	Itamarati II - MST	8.551	529	30/09/2004
48.	Ponta Porâ	New Era	2.802	97	06/11/2000
49.	Ponta Porâ	Boa Vista	2.050	73	07/02/2000
50.	Rio Brilhante	Fortune	2.383	74	11/08/1998
51.	Rio Brilhante	St. Jude	4.155	145	09/10/1998
52.	Rio Brilhante	Fortaleza	384	12	27/10/1999
53.	Rio Brilhante	Margarida Alves	3.166	68	29/12/2000
54.	Rio Brilhante	Taquara	1.550	54	10/08/1998
55.	Rio Brilhante	Triangle	927	41	22/12/1997
56.	Rio Brilhante	Bonsucesso	664	24	27/10/1999
57.	Rio Brilhante	Blue Lagoon	1.507	116	27/12/2005
58.	Rio Brilhante	Silvio Rodrigues	3.202	64	14/09/2001
59.	Sidrolândia	Capào Bonito III	600	22	27/12/2000
60.	Sidrolândia	Terra Solidaria	360	11	28/11/2003
61.	Sidrolândia	Joâo Batista	3.649	3	02/08/2005
62.	Sidrolândia	San Pedro	8.592	292	23/12/1998
63.	Sidrolândia	AltemirTortelli	1.489	108	27/12/2005
64.	Sidrolândia	Alambari FAF	2.809	198	27/12/2005
65.	Sidrolândia	Eldorado Part	834	70	05/12/2007
66.	Sidrolândia	Vista Alegre	1.030	50	10/11/1999
67.	Sidrolândia	Terra Solidaria II	607	18	19/10/2004
68.	Sidrolândia	Barra Nova	3.940	264	27/12/2005
69.	Sidrolândia	Geraldo Garcia	5.775	181	29/12/2000
70.	Sidrolândia	Alambari- CUT	3.126	222	27/12/2005
71.	Sidrolândia	Capào Bonito	2.585	132	20/10/1989
72.	Sidrolândia	Eldorado	9.972	316	15/07/2005

73.	Sidrolândia	Santa Terezinha	1.537	64	13/08/2001
74.	Sidrolândia	boa constrictor	7.218	238	11/07/2000
75.	Sidrolândia	Barra Nova FETAGRI	1.005	1	27/12/2005
76.	Sidrolândia	Capào Bonito II	8.231	308	14/10/1997
77.	Sidrolândia	Nazareth Farm	2.491	169	18/12/2013
78.	Sidrolândia	Eldorado II	9.619	751	27/12/2005
79.	Sidrolândia	Vacaria	1.067	47	27/03/2000
80.	Sidrolândia	Santa Lucia I	1.322	98	09/03/2007
81.	Sidrolândia	Alambari FETAGRI	2.132	143	27/12/2005
82.	Taquarussu	Bela Manhà	1.862	100	11/11/2005

Source: National Institute for Colonization and Agrarian Reform - INCRA. Organization: TEIXEIRA, J. C. - December 2014.

Maps 5, 6, 7, 8 and 9 show land use and occupation in the Ivinhema river basin in 2003, 2006, 2009, 2013 and 2015, demonstrating the great advance of sugarcane cultivation in this area during this period.

The production of maps based on satellite images generated data on land use and occupation in the basin area, which gave us better conditions for analyzing the advance of sugarcane cultivation within the boundaries of this hydrographic basin. The data generated differs from the secondary data in that it considers the entire area of the municipalities, as we have already pointed out.

The land use and occupation map for 2003 (Map 5) shows the greatest concentration of sugarcane plantations in the northwest of the basin. A large part of the basin was occupied by pastures, with extensive livestock farming as the predominant economic activity. Crops were located almost exclusively in the upper part of the basin (northwest, west and southwest), where the most fertile soils are found.

The presence of sugarcane plantations in this part of the basin in 2003 is justified by the fact that it is in the area of the municipalities of Rio Brilhante, Maracaju and Sidrolândia, where there have been mills in operation since the last century. We also observed sugarcane plantations in the area covering the municipality of Nova Andradina, due to the presence of the Santa Helena mill. In addition to these locations, these crops appeared in isolated spots in the basin area.

According to the data generated when the maps were drawn up, in 2003 sugar cane areas occupied 70,509.20 ha, representing 1.52% of the basin's total area. The other crops together occupied 85,515.30 ha, totaling 1.84%. Pastures occupied 3,284,655.24 ha, representing 70.27%; forests 1,036,021.88 ha, totaling 22.16%; exposed soils totaled 78,785.19 ha, or 1.68% of the area.

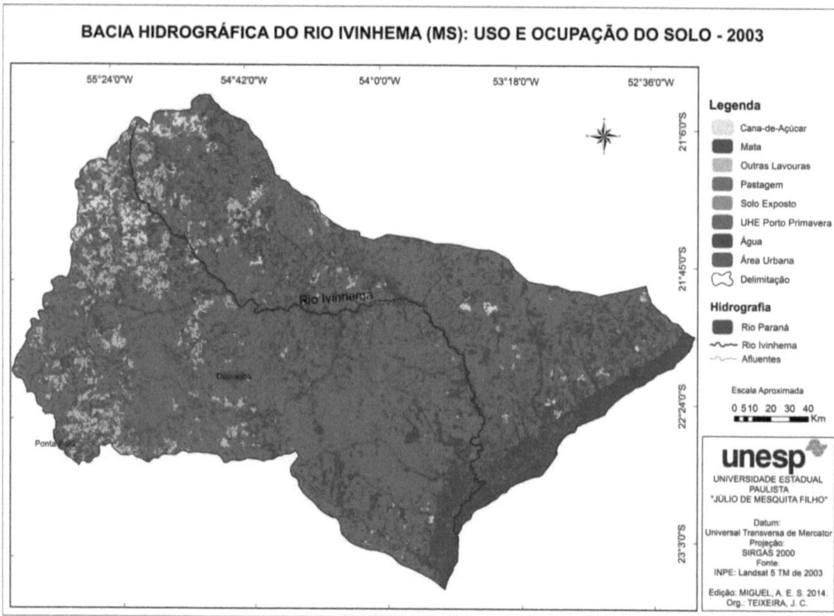

Map 5: Ivinhema river basin (MS): Land use and occupation - 2003.

In 2006, sugar cane areas amounted to 101,026.69 ha, representing 2.18% of the basin's total, while other crops amounted to 164,630.66 ha (3.53%), pasture areas totaled 3,059,078.10 ha (65.44%), woodlands totaled 1,092,853.88 ha (23.38%) and exposed soils 131,722.37 ha (2.81%).

Thus, compared to 2003, there was a slight increase in the areas under sugar cane and other crops, which together accounted for 5.71% of the basin's area, an increase of 2.35 percentage points, while pasture areas fell by 6.9 percentage points. We note the greater advance of sugarcane areas in the north-central direction and in the extreme south of the basin, mainly replacing pasture areas (Map 6).

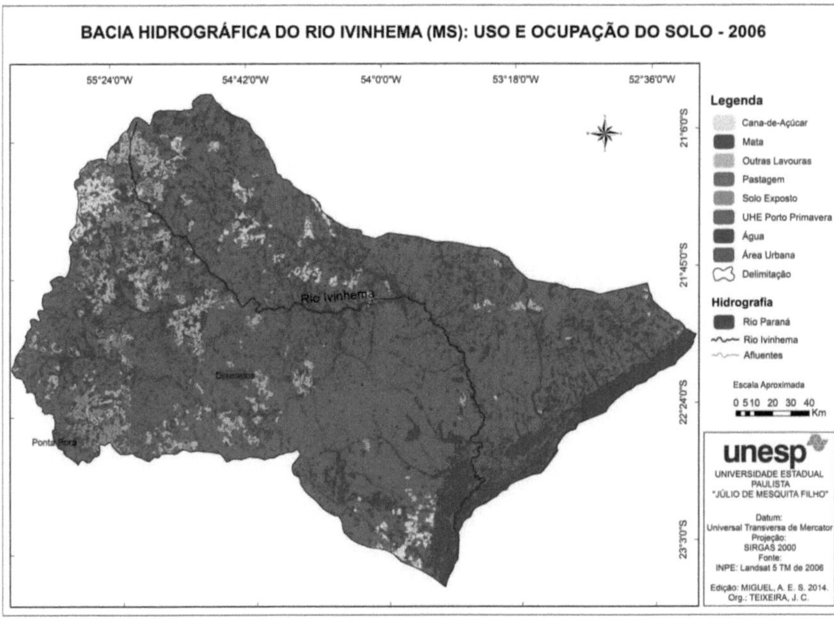

Map 6: Ivinhema river basin (MS): Land use and occupation - 2006.

In 2009, sugarcane areas in the Ivinhema basin totaled 288,462.64 ha (6.18%), while other crops totaled 331,268.82 ha (7.09%), pastures totaled 2,718,678.58 ha (58.17%), forests 1,092,204.01 ha (23.36%) and exposed soils 121,297.61 ha (2.59%).

Therefore, between 2006 and 2009, the previous trend of growth in crop areas, including sugarcane plantations, and a reduction in pasture areas continued. These crops remain heavily concentrated in the upper Ivinhema basin, mainly in the far west, but will also appear in other areas of the basin. Soils that were exposed in 2006 were occupied by crops in 2009, mainly sugar cane (Map 7).

It's important to note that most of the mills began operating during this period, especially in 2008 and 2009, which explains the strong expansion of sugarcane plantations. There were also plans to set up other mills in the area, which did not materialize.

Map 7: Ivinhema river basin (MS): Land use and occupation - 2009.

In 2013, sugar cane areas totaled 920,705.95 ha (19.70%), other crops 526,542.58 ha (11.26%), pastures 2,020,668.29 ha (43.25%), forests 1,041,797.40 ha (22.28%) and exposed soils 35,788.50 ha (0.77%).

Map 8 shows that, in 2013, there was a large expansion of sugarcane production in this geographical area, compared to previous years. However, although the advance of sugarcane production took place throughout the Ivinhema river basin, the greatest concentration of this production continued to be in the highest part of the basin, as well as other crops. Pastures continued to lose ground. This expansion of sugarcane plantations shown on the map reinforces what we saw during our first fieldwork, the high level of land rental for sugarcane production, which has also reduced the area under other crops.

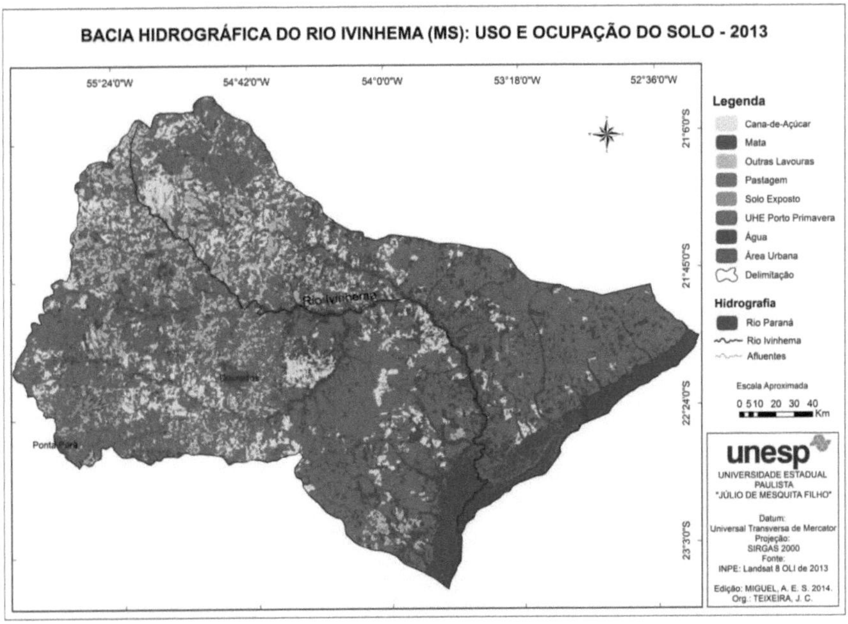

Map 8: Ivinhema river basin (MS): Land use and occupation - 2013.

Map 9 - land use and occupation in 2015 - showed a considerable reduction in sugar cane areas, which now occupy 640,890.10 ha, or 13.71% of the total area of the basin. The areas occupied by other crops continued to grow and reached 895,592.50 ha, representing 19.16% of the total area. Pastureland decreased to 2,000,420.70 ha (42.80%), woodland remained unchanged at 1,010,697.00 ha (21.63%) and exposed soils at 18,601.20 ha (0.40%).

From the data generated in the 2015 land use and occupation map, we can see that the areas occupied by other crops have grown more, once again occupying more space than sugarcane areas. This change was due to the increase in soybean prices and the crisis in the sugarcane sector. In the second stage of the field research, carried out between December 2014 and April 2015, we saw these changes when we observed that some of the sugarcane-producing establishments on the first visit in 2013 had switched to growing soybeans.

The map shows the advance of other crops over the space previously occupied by sugarcane plantations, mainly between the center and the west of the basin area, i.e. in the upper part with the most fertile soils. These crops are represented, in particular, by soya and, to a lesser extent, maize.

On our second visit to the property of Mr. G. A. M, a rural producer in the municipality of Nova Andradina, we observed the switch from sugar cane to soybeans. This producer made the following statement:

> The sugar cane sector is in crisis. Some mills are closing and others are threatening to close. So I went back to soybeans, which today have a good price. In addition to my farm, I've rented two more to plant soybeans and that's the trend over the next few years. Soy will take the place of sugar cane.[32]

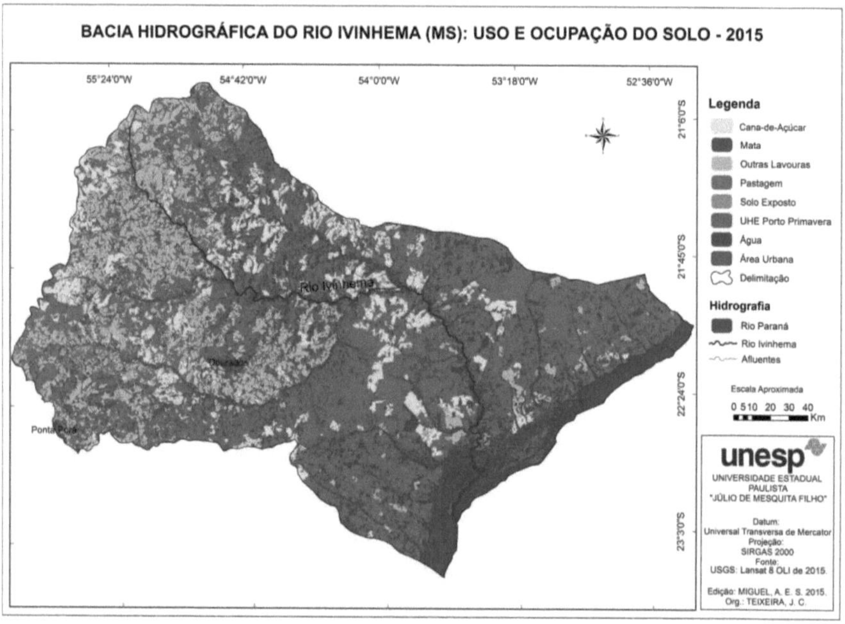

Map 9: Ivinhema river basin (MS): Land use and occupation - 2015.

Comparing the land use and occupation maps for 2003, 2006, 2009, 2013 and 2015, we see a clear reduction in pasture areas and a significant expansion of sugarcane areas, with the exception of 2015, when there was a reduction in its area. This is evidence of what we found in our field research: the advance of sugarcane cultivation has mainly been over pasture areas. This fact is evidenced by the data generated from maps of land use and occupation in this watershed, which show that pasture areas fell from 70.27% in 2003 to 42.80% in 2015, while sugar cane areas expanded from 1.52% to 13.71% in the same period, with a peak in 2013, when they reached 19.70%.

32 Excerpt from an interview conducted in January 2015 with Mr. G. A. M, a rural producer in the municipality of Nova Andradina - MS.

Crop areas have also grown considerably in the basin. In 2003, they occupied 1.84% of the area, but in 2015 they increased to 19.16%. This expansion is represented, in particular, by soya and maize, because as the secondary data shows, there has been a reduction and almost disappearance of other important crops, such as rice and beans. However, this is not directly related to the expansion of sugarcane production, as shown above.

This doesn't mean that sugarcane cultivation hasn't spread over areas of other crops, but it has happened to a lesser extent. There was also the substitution of sugar cane for other crops and the advance of other crops over pasture areas. The geographic space of the Ivinhema River basin has undergone significant changes in the landscape, altering the productive structure rapidly over the last twelve years of analysis.

We also observed a significant area of natural forest that did not undergo many changes over the period, with the percentage of occupation oscillating between 21% and 23%. This could indicate that there was no major deforestation in the basin area during this period. However, we found that in the areas of large crops, trees were uprooted to facilitate the action of harvesting machines. The significant percentage of woodland is due to the fact that this basin covers a reserve area close to the Paranâ River.

Exposed soils appear in all the years analyzed, but do not exceed 3% of the total. These exposed soils are the areas prepared for planting crops, including sugar cane.

Graph 10 illustrates what has been shown so far in the analysis of the maps presented of land use and occupation in the basin area, i.e. the reduction of pasture areas and the expansion of crop areas, including sugar cane.

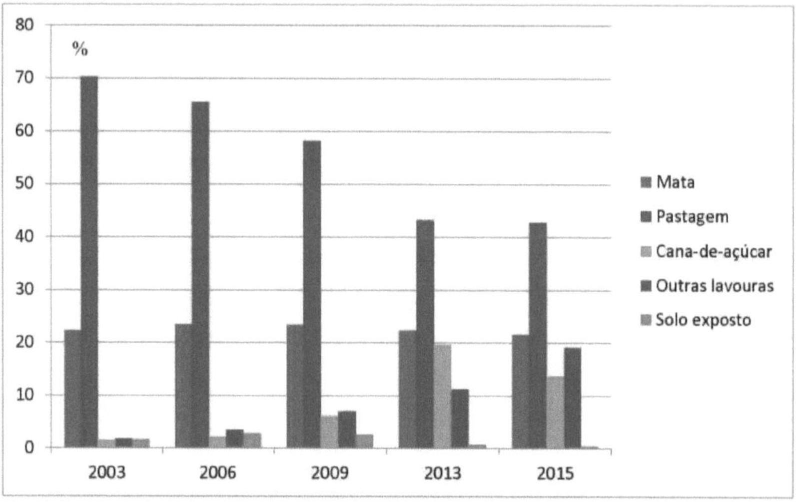

Graph 10: Ivinhema river basin (MS): land use and occupation (2003-2015).

Below is a key interpretation map (Map 10) of the Ivinhema river basin area, based on the use and occupation map from 2015, using points marked with GPS during the fieldwork.

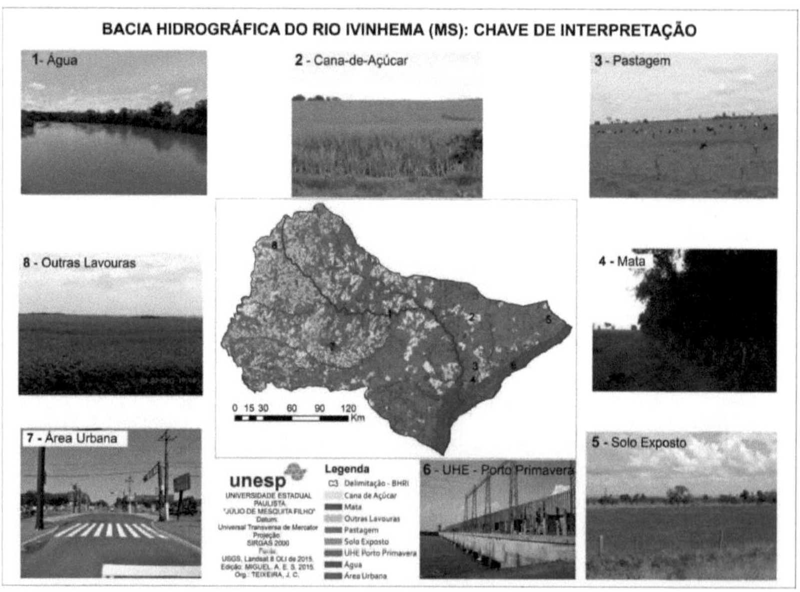

Map 10: Ivinhema River Basin (MS): Interpretation Key.

The data shows that other crops resumed growth in the year 2015, which was confirmed in the fieldwork, given the expansion of areas planted with soya. The rise in the price of a sack of soybeans and the crisis in the sugarcane sector explain this.

The effects of the crisis in the sugar-alcohol sector in recent years have also been reflected in the performance of the sector's mills in the basin area. These mills are struggling to stay in operation. Of the 17 plants in operation in 2014, two stopped working, CBAA in Sidrolândia (in 2014) and Aurora in Anaurilândia (in 2015), shown in figures 15 and 16. Many others are currently in a delicate situation.

Figure 15: Anaurilândia - MS: Aurora Mill.

Source: Fieldwork (Photo by the author - December 2014).

Figure 16: Sidrolândia - MS: Usina CBAA- Sidrolândia.

Source: Fieldwork (Photo by the author)

The effects of the crisis have been announced by several companies in the sector. According to the Union of Bioenergy Producers (UDOP), the sector's discouragement is related to the competition between ethanol, the main product, and gasoline subsidized by the federal government, especially since the discovery of pre-salt oil.

According to data from the Brazilian Sugarcane Industry Association (UNICA), there is a tendency for several companies in the sector to close between 2014 and 2016 in the Center-South.

The Biosev group announced a net loss of over a billion reais in the last harvest. Bunge has announced that it will cease sugar production in Brazil in the face of falling prices on the international market. The Odebrecht group also reported layoffs in 2014 and plans to shut down some mills from 2015 if the situation doesn't improve. The Sao Fernando mill, the state's largest producer of electricity through cogeneration by burning sugarcane biomass, went into receivership in 2013, in order to pay off its debts to banks, especially the BNDES, labor debts and debts to suppliers (BATISTA, 2014).

Given the current trend, the following questions remain: what is the future of the sugar-alcohol sector in the region? Will it be able to overcome the current crisis and resume the growth seen in the 2000s? Or will there be a new restructuring of agricultural production in the basin area in the face of the closure of other mills? These are questions suggested for future research.

And while we started the project by hypothesizing that food production areas were being replaced by sugar cane in the Ivinhema River basin area, we ended up with the thesis that these areas of sugar cane plantations have become much more pastureland, reducing livestock farming in the region. In addition, there was also an expansion of soybean and corn areas over pasture areas and even the replacement of crops by other crops.

Even with the metamorphosis of this regional agricultural space, it seems to us that agriculture is tending to grow stronger, albeit in the interests of agribusiness, be it sugar cane, soybeans or corn. And within this space, family farming resists, as it hasn't entered the fray of the big agricultural crops and continues to produce foodstuffs, especially in the rural settlements in this area.

FINAL CONSIDERATIONS

In the first decade of the current century, there was a major expansion of the sugar-alcohol sector in Brazil, leading to the expansion of sugarcane plantations into the interior of the country and the installation of new mills and distilleries.

With this in mind, the sugarcane agribusiness entered Mato Grosso do Sul and there was a significant expansion of sugarcane plantations, occupying a large part of its agricultural space, and several sugar and alcohol mills were set up in the state.

In this context, the area of the Ivinhema river basin has come to concentrate most of the state's sugarcane production, with the municipalities of Rio Brilhante, Nova Alvorada do Sul, Dourados Maracaju, Angélica, Nova Andradina and Ponta Porâ standing out as having the highest production in recent years.

Between 2003 and 2013, this production grew above the national average, showing the interest of the sector's entrepreneurs in this area of Mato Grosso do Sul. During this period alone, 13 mills were set up in the basin area and, despite a reduction in the space occupied by sugarcane plantations in the last two years, it currently accounts for almost 80% of the state's sugarcane production.

The area of the Ivinhema River basin has a favorable climate for growing sugar cane, the relief is not very undulating, the soils are fertile, there is an available workforce and the state and municipal governments have offered facilities with the aim of attracting companies linked to the sugar-alcohol sector. Despite its attractions, the region faces difficulties related to infrastructure, especially internal logistics and the distance from ports.

The sugarcane plantations expanded through the leasing and purchase of land by the mill owners, and the number of producers who started producing sugarcane on their own and selling it to these companies increased.

In the mid-2000s, leases for sugar cane cultivation became easier to sign, as producers saw it as an alternative for obtaining higher yields in the face of falling soybean prices and difficulties with livestock farming.

In the Ivinhema river basin, sugar cane has also started to share space with other crops and livestock, as some producers have chosen not to completely abandon their previous activity, even planting sugar cane on part of their farms. Thus, in this area it is common to find sugar cane sharing space with soybeans, corn, livestock and even manioc.

The greatest concentration of sugarcane plantations occurs in the upper part of the basin, where the most fertile soils are found, occupying spaces that could be producing basic foodstuffs.

At first, given the reduction in the area under staple food crops, as shown by the secondary data, we hypothesized that these crops would be replaced by sugar cane, but this was not confirmed by the fieldwork, because these products are grown by small producers and they have shown no interest in sugarcane production due to the small size of their farms and, for the same reason, the mills have no interest in acquiring or leasing these areas.

The areas of fertile soil used for other crops were occupied by soybeans and corn and not by other foodstuffs, as we assumed at the beginning of the research. In the period analyzed, these always occupied the small rural establishments. Their reduction was due to small producers switching to other crops in search of alternative income or abandoning production and looking for jobs outside their establishments.

The presence of small producers in the basin area is significant and has increased since the establishment of rural settlements, which partly guarantees the production of some products that are not of interest to large producers.

These small producers face difficulties in maintaining production, despite the fact that some use PRONAF resources and participate in PNAE. Many of them use their farms only as a home and to grow products for their own families. In this case, these producers work in the cities, in the mills or in other rural establishments.

We noticed that, even in the face of legislation, the mills often don't respect the limits of these establishments when spraying from the air, which has led to conflicts in the region. Other effects resulting from the expansion of sugarcane plantations are the removal of trees to facilitate mechanical harvesting, as well as burning which, despite the ban and constant reduction in recent years, still occurs in isolated spots. We cannot underestimate this fact, even if this approach is not part of our objectives, because we know that nature has become of secondary importance to the interests of agribusiness.

Other conflicts include the difficulties in demarcating indigenous lands and the presence of landless people in the face of agribusiness, occupying the sides of highways.

Thus, we can conclude that the expansion of the sugarcane sector has generated various socio-spatial effects over a twelve-year period, causing a real metamorphosis of agricultural space and changing the landscape.

The advance of sugarcane plantations in the basin area was mainly over pasture areas, but also over areas of other crops. Most of the crops that gave way to sugar cane were grains (soybeans and corn), also grown by larger producers.

Despite the expansion of other crops, sugar cane was the one that most entered the pasture areas. In this way, we can say that there has been a spatial rearrangement, with the territorialization of sugarcane monoculture causing the deterritorialization of livestock.

The establishment of sugar and ethanol mills has created new jobs, but it hasn't always absorbed only workers from these municipalities, as specialized labor usually comes from other regions. Temporary jobs are the most significant, especially during the harvest period, but with increasing mechanization, this is changing and fewer people are being hired. Among the workers hired are small producers who seek sources of income outside their properties to supplement the family budget. We were also informed that indigenous people usually work in the manual harvesting of sugar cane, but we didn't find any cases when we carried out our field research. However, the biggest generation of jobs took place when these companies were set up, employing people both in the internal works and in preparing the soil and planting the crops. We know that job creation indirectly influences trade, as it means more money circulates, so this can be considered a positive aspect.

While jobs have been generated on the one hand, on the other, it has also contributed to the closure of jobs in other sectors. The reduction in the number of cattle, for example, has resulted in the closure of meatpacking plants, which means fewer jobs. The hiring of workers who were previously dedicated to family farming also has an impact on labor relations, since they could dedicate themselves to working in their establishments.

The events of the 2000s indicated a promising future for the sugar-alcohol sector in this area of the state. However, the current crisis in the sector has led to changes in the production process in the basin area. New facts have emerged that raise questions about the future of the sector in the region. These include the financial difficulties and indebtedness of the mills, including the closure of two of them in the area; the failure to implement the new projects for this sector that were presented at the end of the 2000s; the dismissal of employees from the mills, especially in 2014 and 2015; the reduction in sugarcane plantations and the constant lack of interest in continuing with this activity, recently demonstrated by producers; the high prices of soybeans, causing the expansion of this crop to resume. Therefore, strategic planning is needed to rebalance the sector, bringing together businesspeople and the government.

The part of the state where the Ivinhema river basin is located has always been the area with the highest agricultural production in the state and, even with the changes that have occurred with the sugar cane *boom* in the last twelve years, it has maintained its position as a major producer, mainly of grain. The presence of family farming has also always been important in the production of basic foodstuffs, including milk, despite the difficulties faced by the sector.

The guarantee of maintaining the production of these products, which are fundamental for families' livelihoods as well as for supplying urban markets, does not lie in preventing the advance of sugarcane and other crops for export, but rather in giving greater support to small producers, because even though they are the most important in the production of these foods, they remain on the margins of the production process and do not dedicate themselves as they should to family farming activities, including employment in cities, mills and large rural establishments.

We have to consider that the economy of most of the municipalities in the basin continues to depend on good results from agriculture, bearing in mind that, for them, industrialization is not yet the main generator of income.

The future of the sugar-alcohol sector in this region will depend on the actions of the government and the businessmen involved, as well as the direction of the market for sugarcane products. In this sense, we affirm that the Ivinhema River basin area will continue to play an important role as an agricultural producer in the state, and that the changes in this area signal changes in crops, which does not mean a reduction in plantations. Despite the presence of small producers in this area, the major dominance of the productive space has been exercised by corporate producers, which means that if there are no structural changes that alter the land structure and value family farming, the tendency is for agribusiness to continue to dictate the rules of this important agricultural space in the state of Mato Grosso.

In view of the above, we believe we have achieved our main objective: to analyze sugarcane expansion in the Ivinhema river basin over the last twelve years and its effects on this part of the state of Mato Grosso do Sul.

The procedures adopted were sufficient to achieve this goal. However, we experienced some difficulties during the course of the research, especially related to the large territorial extent of the study area, requiring extensive fieldwork to understand its agricultural production dynamics. The biggest difficulty we encountered was gathering data on the mills, as these companies showed no interest in providing the information we were looking for. Only five of the seventeen companies located in the area contacted us, and even then they didn't provide all the information requested in the survey form.

The socio-spatial approach, using the area of a river basin as the spatial cut-off, was the challenge we set out to tackle from the start of the research, even though we were aware of the difficulties we would encounter, especially in analyzing secondary data, because it is normally systematized for political and administrative spaces. Even so, we believe that this experience has been a positive one, as it has shown us that it is possible to approach a space with imprecise boundaries, such as river basins.

Thus, we hope to have contributed to an understanding of the productive dynamics of this important agricultural area in the state of Mato Grosso do Sul, and that this research can add qualitatively to other analyses of the progress of the sugar-alcohol sector in the country. Finally, that it may raise other questions and arouse interest in carrying out further research in the Ivinhema river basin.

REFERENCES

ABRAMOVAY, R. A. Actualidade do Mètodo Josué de Castro e a Situaçâo Alimentar Mundial. In: **Revista de Economia e Sociologia Rural.** vol. 3 e 4, n° 1, jul/dez 1996, p. 81-102.

ALVES, G. **O novo (e precàrio) mundo do trabalho:** Reestruturaçâo produtiva e crise do sindicalismo. Sao Paulo: Boitempo, 2000.

ANDRADE, E. T. de; CARVALHO, S. R. G. de; SOUZA, L. F. de. PROÀLCOOL PROGRAM AND ETHANOL IN BRAZIL. In: **ENGEVISTA**, V. 11, n. 2, December/2009, p. 127-136.

ANDRADE, M. C. de. **History of the Sugar Mills of Pernambuco**. Recife: Editora Massangana - Joaquim Nabuco Foundation, 1989.

Area of the sugarcane system. Recife: Sudene, 1988.

Modernization and Poverty: the expansion of the Sugarcane Agroindustry and its ecological and social impact. Sao Paulo: Editora da Unesp, 1994.

ARBEX JÙNIOR, J; OLIC, N. B. **Rumo ao Centro-Oeste:** o Brasil em regioes. Sâo Paulo: Editora Moderna, 1996.

ARRUZZO, R. C. Agricultural modernization, work and spatial organization on the BR-163. In: BERNARDES, J. A; FREIRE FILHO, O. L. **Geografia da Soja:** BR 163, fronteiras em mutaçâo. Rio de Janeiro: Arquimedes, 2005, p. 99-114.

ASEVEDO, T. R. A. de; RIBEIRO, J. C. Capital, labor and the State in the expansion and strengthening of sugarcane agribusiness in Mato Grosso do Sul: some intertwining of recent antagonisms. In**: Anais do XVI Encontro Nacional de Geógrafos - XVI ENG, crise, pràxis e autonomia: espaços de resistências e esperanças, espaço de diàlogos e pràticas**. Porto Alegre, 2010, ISBN 978-85-99907-02-3.

AZEVEDO, J. R. N. de. Perspectives on land ownership in Mato Grosso do Sul. In: **Revista Entre-Lugar**, Dourados, MS, ano 1, n. 1, 2010, p. 49-71.

BACCARIN, J. G. **A constitçâo da Nova Regulamentaçâo Sucroalcooleira.** Sâo Paulo: Editora UNESP. 2005. 237 p.

BACKES, T. R. From the homogenization of the landscape to the expanded reproduction of capital: an analysis of the expansion of sugarcane in the state of Mato Grosso do Sul. In: **Caminhos de Geografia,** Uberlândia v. 9, n. 26 Jun/2008 p. 173 -

184. Available at: http://www.ig.ufu.br/revista/caminhos.html. ISSN 1678-6343.

BALSAM, R. Impacts of the modernization of Brazilian agriculture. In: **Campo Territòrio: revista de geografia agrària**, v. 1, n. 2, p. 123-151, aug. 2006.

BATISTA, F. Biosev has a net loss of R$ 1.017 billion in the 4th quarter of 2013/2014. In: **Valor Econòmico.** Wednesday 11.6.2014. Available at: http://www.novacana.com/n/indùstria/financeiro/bioserv-preiuizo-liquido- 110614. Accessed on: Jul. 2014.

BORGES, U. FREITA, G. H. HURTIENNE, T. NITSCH, M. **PROALCOOL:** economics, politics and socio-economic evaluation of the Brazilian biofuel program. Aracajù: UFS, 1988.

BRAZIL - WWF - AGRICULTURE AND ENVIRONMENT PROGRAM. **Analysis of the expansion of the sugarcane agro-industrial complex in Brazil**.

2008. Available at:< http://www.ambiente.sp.gov.br/wp- content/uploads/.../rel cana wwf.pdf> . Accessed on: Oct. 2013.

BRAZIL. BRAZILIAN AGRICULTURAL RESEARCH COMPANY - EMBRAPA. **Agro-ecological zoning of sugar cane.** Brasilia: EMBRAPA, 2009.

BRAZIL. Brazilian Institute of Geography andStatistics (IBGE). **Agricultural census of Mato Grosso do Sul 1995/1996.** Available at:< http://www.ibge.gov.br> . Accessed on: Mar. 2013.

Brazilian Institute of Geography andStatistics (IBGE). **Agricultural census of Mato Grosso do Sul in 2006.** Available at< http://www.ibge.gov.br> . Accessed on: Mar. 2013.

Brazilian Institute of Geography and Statistics (IBGE). **1975 Agricultural Census of Mato Grosso do Sul.** Rio de Janeiro: IBGE, 1975.

Brazilian Institute of Geography and Statistics (IBGE). **1985 Agricultural Census of Mato Grosso do Sul.** Rio de Janeiro: IBGE, 1985.

Brazilian Institute of Geography and Statistics (IBGE). **2010 Demographic Census.** Available at< http://www.ibge.gov.br> . Accessed on: Mar. 2013.

Brazilian Institute of Geography and Statistics (IBGE). **Municipal Agricultural Production 2004**. Available at< http://www.ibge.gov.br> . Accessed on: Mar. 2013.

Brazilian Institute of Geography and Statistics (IBGE). **Municipal Agricultural**

Production 2006. Available at< http://www.ibge.gov.br> . Accessed on: Mar. 2013.

Brazilian Institute of Geography and Statistics (IBGE). **Municipal Agricultural Production 2009**. Available at< http://www.ibge.gov.br> . Accessed on: Mar. 2013.

Brazilian Institute of Geography and Statistics (IBGE). **Municipal Agricultural Production 2012**. Available at< http://www.ibge.gov.br> . Accessed on: Mar. 2013.

BRAZIL. Ministry of Agriculture, Livestock and Supply (MAPA). **Statistics and basic data on agricultural economics**. Available at http://www.agricultura.gov.br. Accessed on: Nov. 2013.

BRAY, S. C; FERREIRA. E. R; RUAS, D. G. G. **As politicas da agroindùstria canavieira e o PROÀLCOOL no Brasil.** Marilia: Unesp-Marilia Publicaçôes, 2000.

BRITO, A. BNDES releases R$5.2 billion for sugarcane mills this year. In: **Folha de Sâo Paulo newspaper**. São Paulo, November 19, 2008.

BRUM, A. J. **Modernização da agricultura:** trigo e soja. Petrópolis: Vozes, 1988.

CAMARGO, A. M. M. P. de; CASER, D. V; CAMARGO, F. P. de; OLIVETTE, M. P. A; SACHS, R. C. C; TORQUATO, S. A. Dynamics and trend of sugarcane expansion over other agricultural activities, State of São Paulo, 2001-2006. In: **Revista Informaçôes Econômicas**, SP, v.38, n.3, mar. 2008.

CAMPESTRINI, H; GUIMARAES, A. V. **História de MS**. Campo Grande: Gràfica e Papelaria Brasilia, 1991.

CARONE, E. **The Old Republic**. Sâo Paulo: Difel, 1970.

CARRIJO, E. L. O; MIZIARA, F. A expansâo do setor sucroalcooleiro como uma nova etapa **da** fronteira agricola em Goiàs: estudo de caso no município **de** Mineiros In: **Revista de Economia da UEG**. Anâpolis: UEG, vol. 05, n° 02, JUL-DEZ/2009, p. 92-121.

CARVALHO, R. G; SANTOS, J. C. dos. The process of sugarcane expansion in the geographic micro-region of Ituiutaba (MG) and the degradation of the sugarcane worker. In: **Proceedings of the XXI National Meeting of Agricultural Geography - territories in dispute: the challenges of Agricultural Geography in the contradictions of Brazilian development**. Uberlândia: UFU, 2012.

CARVALHO, S. P. de; CARRIJO, E. L. O. Alcohol production: from Proàlcool to the current context. In: **Proceedings of the XLV Congress of the Brazilian Society of Economics, Administration and Rural Society.** Londrina: UEL, 2007.

CASTRO, E. R. de. **Rural credit and agricultural supply in Brazil**. Thesis (PhD in Applied Economics). Viçosa: UFV, 2008.

CASTRO, F. C. **Family farming in Brazil**: new challenges for the 21st century. 2003. Available at: http://www.geoufsc.com.br. Accessed on: Jul. 2013.

CASTRO, S. S. de; ABDALA K; SILVA, A. A; BORGES, V. The expansion of sugar cane in the Cerrado and in the State of Goiás: elements for a spatial analysis of the process. In: **Boletim Goiano de Geografia**, 2010.

CLEPS JÙNIOR, J. The agrarian question, the State and territory in dispute: approaches to agribusiness and the nature of conflicts in the Brazilian countryside. In: SAQUET, M.A; SANTOS, R.A. (Org.) **Geografia agrària, territòrio e desenvolvimento.** Sâo Paulo: Expressâo Popular, 2010. p. 35-54.

COELHO, C. N. 70 anos de política agricola no Brasil (1931-2001). In: **Revista de Politica Agricola**, Brasilia, n.3, p.3-58, jul./aug./sept., 2001. (Special issue 2001).

CORDEIRO, A. Ethanol to feed cars or food to feed people? In: **Impacts of the sugarcane industry in Brazil**. 2008, p. 9-22. Available at: http://www.plataformabnds.org.br. Accessed on: October 2012.

COSTA, R. C. **History of sugar cane in Brazil**. 2008. Available at http://www.scgms.com.br. Accessed June 2013.

COSTA, R. M. da; CHRYSOSTHEMOS, R. N; ALVES, F. J. da C. Historical and political aspects of the expansion of sugarcane agro-industrial centers in the state of Mato Grosso, Brazil. In: **Proceedings of the 48th SOBER - Brazilian Society of Economics, Administration and Rural Sociology**. Campo Grande, 2010.

CUNHA, J. M. P. da. **Migratory dynamics and the process of occupation of the Brazilian Midwest:** the case of Mato Grosso. Campinas: Nùcleo de Estudos de Populaçâo / Unicamp, 2011, 87 p.

D'ALKMIN FILHO, P; TEIXEIRA, J. C. Family farming in school meals in Nova Andradina, MS. In: **Proceedings of the XXI National Meeting of Agrarian Geography. Territories in dispute: the challenges of Agrarian Geography in the contradictions of Brazilian development**. Uberlândia: UFU, 2012, ISSN 1983-487X.

DELGADO, G. da C. **Capital financeiro e agricultura no Brasil**: 1965-1985. Sâo Paulo: icone: Campinas, UNICAMP. 1985.

DUARTE, A. C. The Midwest in the Regional Organization of Brazil. In: **Geografia do**

Brasil - Regiâo Centro-Oeste. Rio de Janeiro: IBGE, 1989, p.15-20.

FARIA, A; FRATA, A. **Biofuels:** sugarcane in the Paranà river region and grain production, livestock and sugarcane in the Ivinhema river sub-basin. Campo Grande: Ecoa, 2008.

FARIAS, C. S. de; SILVA, S. S. da. The implications of sugarcane agribusiness for food production in the south-western Amazon. In: **Anais do XXI Encontro Nacional de Geografia Agrària - territórios em disputa: os desafios da Geografia Agrària nas contradiçôes do desenvolvimento brasileiro**. Uberlândia: UFU, 2012.

FAUCONNIER, R; BASSEREAU, D. **La canne à sucre**. Paris: Maisonneuve & Larose, 1970.

FERNANDES, B. M. The new name is agribusines. **Nera Publications**, 2004. Available at: http://www4.fct.unesp.br/nera/publicacoes/onomeeagribusiness.pdf. Accessed on: Nov. 2013.

FERREIRA, F. J; BRANDÂO, E. C. Educaçâo do campo: um olhar histórico, uma realidade concreta. In: **Revista Eletrônica de Educaçâo.** Year V. No. 09, July/December 2011.

FRATA, A. M; FARIA, A. B. de. The expansion of sugar cane and the cut-off in hydrographic basins: ethanol in the Ivinhema river sub-basin. In: **Proceedings of the 48th SOBER Congress - Brazilian Society of Economics, Administration and Rural Sociology.** Campo Grande, July 2010.

FURTADO, C. **Formaçâo Econômica do Brasil**. Rio de Janeiro: Fundo de Cultura, 1961.

GERARDI, L. H. de O. Some reflections on the modernization of agriculture. In: **Geografia**, Rio Claro, v. 5, n. 9/10, 1980, p. 19-34.

GOES, T; MARRA, R. **The expansion of sugarcane and its sustainability**. Embrapa, 2009. Available at: http://www.agencia.cnptia.embrapa.br. Accessed on: Jan. 2014.

GONÇALVES NETO, W. **Estado e agricultura no Brasil**: politica agricola e modernizaçâo econômica brasileira, 1960-1980. Sâo Paulo. Hucltec, 1997.

GRAZIANO DA SILVA, J. **A nova dinâmica da agricultura brasileira.** Campinas: UNICAMP/IE, 1996.

The **new Brazilian rural world**. Campinas: Unicamp, 2000. 151 p. Research series.

The New Brazilian Countryside. In: SHIKI, S; GRAZIANO DA SILVA, J; ORTEGA, A. C. **Agricultura, Meio Ambiente e Sustentabilidade do Cerrado Brasileiro**. Uberlândia: UFU, 1997, p. 75-100.

Technical progress and labor relations in agriculture. Sao Paulo: Hucitec, 1981. 210 p. (Economia & Planejamento. Série "Teses e Pesquisas").

A modernização dolorosa: estrutura agrària, fronteira agricola e trabalhadores rurais no Brasil. Rio de Janeiro: Zahar, 1981.

GRAZIANO NETO, F. **A questão agrària e ecologia**: critica da moderna agricultura. Sao Paulo: Brasiliense, 1985. 154 p.

GUANZIROLI, C. E. Agribusiness in Brazil: prospects and limitations. In: **Economia. Text for discussion**, April/2006. Available at: http://www.uff.br/econ. Accessed on: Mar. 2013.

GUIMARAES, A. P. **A crise agrària**. Rio de Janeiro: Paz e Terra, 1979.

GUIMARAES, E. N; LEME, H. J. de C. **Historical Characterization and Spatial Configuration of the Productive Structure of the Midwest**. 1997. Available at:

http://www.nepo.unicamp.br/textos publish/pronex/pronexbook03/03pronex.

Accessed on: June 2013.

HESPANHOL, A. N. Bioenergy and food crops: producing for whom? In: **Revista Formaçao**, n.15, volume 2, Presidente Prudente: Unesp, 2008a - p.189-226.

Modernization of agriculture and territorial development. In: **4th NATIONAL MEETING OF RESEARCH GROUPS - ENGRUP**, Sao Paulo, 2008b, p. 370-392.

Agribusiness and the spatial reconfiguration of the main crops in Brazil (1990-2006) In: **Revista Geonordeste**. Special Issue, Year XIX, No. 1, Sâo Cristóvam: Federal University of Sergipe, 2008, p. 63-86.

The development of the countryside in Brazil. In: FERNANDES, B. M; MARQUES, M. I. M; SUZUKI, J. C. **Geografia Agrària, teoria e poder**. Sao Paulo: Expressâo Popular, 2007, p. 271-287.

KAGEYAMA, A. *et. al.* **O Novo Padrao Agricola Brasileiro**: do complexo rural aos complexos agro-industriais. Campinas: S/N, 1987.

KUDLAVICZ, M. **Agrarian dynamics and the territorialization of the cellulose/papel complex in the micro-region of Três Lagoas/Ms.** Dissertation

(Masters in Geography), Três Lagoas: CPTL/UFMS, 2011.

LIMA, A. G. The hydrographic basin as a field of study in Human Geography. In: **Revista Geografia** - v. 14, n. 2, jul./dez. 2005 - Universidade Estadual de Londrina, Departamento de Geociências.

MAGALHAES, J. P. de A; KUPERMAN, N; MACHADO, R. C. **Proàlcool**: uma avaliaçâo global. Rio de Janeiro: ASTEL, 1991.

MARAFON, G. J. **Industrialização da Agricultura e Formaçâo do Complexo Agroindustrial no Brasil**. Rio de Janeiro: Geo UERJ N◦ 3, 1998, p. 7-21.

MATO GROSSO DO SUL. SECRETARY OF STATE FOR THE ENVIRONMENT, PLANNING, SCIENCE AND TECHNOLOGY (SEMAC). **Socioeconomic Diagnosis of Mato Grosso do Sul**. 2009. Available at: http://www.semac.ms.gov.br. Accessed on: Aug. 2013.

STATE SECRETARIAT FOR THE ENVIRONMENT, PLANNING, SCIENCE AND TECHNOLOGY (SEMAC). **Socio-economic diagnosis of Mato Grosso do Sul**. 2011. Available at: http://www.semac.ms.gov.br/control.php?id=33720. Accessed on: Aug. 2013.

MATOS, P. F; PESSÔA, V. L. S. The modernization of agriculture in Brazil and the new uses of the territory. In: **Revista Geo UErJ** - Ano 13, n°. 22, v. 2, 2011, p. 290-322 - ISSN 1981-9021.

MENDONÇA, M. R. **The spatial warp of capital and labor in the Cerrado of southeastern Goiás.** 2004. 448 f. Thesis (Doctorate in Geography), Presidente Prudente: FCT-UNESP, 2004.

BIOENERGY AND THE VIABILITY OF FOOD PRODUCTION: for whom? In: **Revista Formaçâo**, n.15, volume 2, Presidente Prudente: Unesp, 2008 -p. 215-226.

Agrobioenergy: the myths of the sugar-alcohol sector in Goiás. In: **Proceedings of the National Meeting of Agricultural Geography**. Sâo Paulo: AGB-USP, 2009, p. 1-22.

MENEZES, A. P. Colònia Agricola Nacional de Dourados (CAND): the work of migrants and the intensification of agriculture in the former south of Mato Grosso. In: **Proceedings of the IV International Conference on Economic History. VI Economic History Postgraduate Meeting**. Sâo Paulo: USP, 2012.

MICHELLON, E, *et. al.* Brief description of Proàlcool and future prospects for ethanol produced in Brazil. In: **Proceedings of the XLVI Congress of the Brazilian Society**

of Economics, Administration and Rural Sociology - SOBER, Rio Branco, 2008,

MIZUSAKI, M. Y. **Territory and productive restructuring in poultry farming**. Dourados: Editora da UFGD, 2009. 356 p.

MONTEIRO, C.V. **Sugarcane in Goiás and Mato Grosso do Sul**. 2012, (text for discussion). Available at: http://www.scgms.com.br. Accessed on: Aug. 2014.

Impacts of the advance of sugarcane in the Midwest. 2010, (text for discussion). Available at: http://www.scgms.com.br. Accessed on: Jul. 2013.

Mato Grosso do Sul in the face of sugarcane expansion. 2014, (text for discussion). Available at: http://www.scgms.com.br. Accessed on: Aug. 2014.

Sugarcane in the Midwest. 2006, (text for discussion). Available at: http://www.scgms.com.br Accessed on: Jul. 2013.

MORAES, R. O. **The Brazilian countryside at the beginning of the 21st century.** Belém: scbc, 2011, 197 p.

MOTA, F. L; PESSÔA, V. L. S. Agribusiness as (re) producer of a new territory: Balsas in the context of soybean agribusiness. In: **Proceedings of the 5th International Symposium on Agricultural Geography - V SINGA.** Rio de Janeiro: UFF, 2009.

MÜLLER, G. **O Complexo Agroindustrial**. Sao Paulo: FGV, 1989.

NASSAR, A. M; RUDORFF, L. B. A.; AGUIAR, D. A.; BACCHI, M. R .P; ADAMI, M. Prospects of the sugarcane expansion in Brazil: impacts on direct and indirect land use changes. In: ZUURBIER, P; VOOREN, J. V. (Edit) **Sugarcane Ethanol:** contributions to climate change mitigation and the environment. 1st Wageninguen: Ed. Wageniguen Publs, 2008, p. 63-94.

OLIVEIRA, A. M. S. de. The new frontiers of sugarcane agribusiness in Brazil: weaving readings on working conditions and food security. In: **Revista Pegada**. Presidente Prudente: UNESP, vol. 8, n.1, 2007. Available at:< http://www4.fct.unesp.br/ceget/PEGADA91/02-9-1-ThomazJr.pdf>. Accessed on: Jun. 2012.

Territorial and productive reorganization of sugarcane agribusiness in Brazil and the consequences for work. Thesis (PhD in Geography), Presidente Prudente: FCT/UNESP, 2009.

OLIVEIRA, A. U. de. Agriculture and industry in Brazil. In: **CAMPO- TERRITÓRIO**: revista de geografia agrària, v.5, n.10, aug. 2010, p. 5-64.

Agrofuels and food production. 2008. Available at: http://www.mst.org.br. Accessed on: Apr. 2012.

Barbarism and modernity: transformations in the countryside and agribusiness in Brazil. In: **Terra Livre magazine**. n. 21. Sao Paulo: AGB, 2003.

Land use, social movements and transformations in the countryside. In: FEITOSA, A. M. A; ZUBA, J. A. G.; CLEPS JUNIOR, J. (Org.). **Debaixo da lona:** tendências e desafios regionais da luta pela posse da terra e reforma agrària no Brasil. Goiânia: UCG, 2006. p. 29-78.

OLIVEIRA, H; URCHEI, M. A.; FIETZ, C. R. **Physical and socioeconomic aspects of the Ivinhema river basin**. Dourados: Embrapa Agropecuària Oeste, 2000.

OLIVEIRA, T. C. M. de. **Agroindustry and the Reproduction of Space:** The Case of Soy in Mato Grosso **do** Sul. Thesis (doctorate in economics), Sao Paulo: USP/FFLCH, 1993. PEREIRA, M. C; MICHELS, I. L; RODRIGUES, J. D; CAMPELO, E. H. R; MEURER, R. A expansao da cadeia sucroalcooleira em Mato Grosso do Sul. In: **Proceedings of the XLV Congress of the Brazilian Society of Economics, Administration and Rural Sociology**. Londrina: UEL, 2007.

PINA, H. **The sugar industry and its legislation**. Sao Paulo: Apec Editora, 1972.

PRADO JÙNIOR, C. **História Econòmica do Brasil**. Sao Paulo: Brasiliense, 35ª Ed., 1987.

REZENDE, G. C. **Agricultural occupation and agrarian structure in the Cerrado**: the role of land prices, natural resources and technology. Rio de Janeiro: IPEA, 2002.

RODRIGUES, J. B. **Isto é MS.** Sao Paulo: Vagner Bicego, 1978.

ROSA, P. A. da. **Historical review of MT**: border with Paraguay. Sao Paulo: S.L.P, 1962.

SANTANA, E. P. de. **Rural Extension in the State of Goiás**: Family production's access to modernity (1975 - 1999). Goiânia: UFG, 2005.

SANTOS, C. S. de A. **The settlers and the Catholic Church in the context of the National Agricultural Colony of Dourados (1940-1970)**. 2007. Dissertation (master's degree in History). UFGD. Dourados.

SANTOS, M. **A natureza do espaço**: técnica e tempo, razao e emoçao. Sao Paulo: Hucitec, 1994.

Space and society: Essays. 2ª ed. Petrópolis: Vozes, 1982.

SILVEIRA, M. L. **O Brasil - territòrio e sociedade no inicio do século XXI.** Rio de Janeiro: Record, 2001.

SANTOS, R. F; SCHLINDWEIN, M. M. Anâlise do nível de desenvolvimento socioeconômico do Estado de Mato Grosso do Sul. In: **Revista História em Reflexao,** vol. 5, n. 9, Dourados: UFGD, Jan/Jun 2011.

SCHNEIDER, S. **Pluriactivity in family farming**. Porto Alegre: UFRGS, 2003.

SEGALLA, A. L. **Botany, improvement and varieties, cultivation and fertilization of sugar cane**. Sâo Paulo: Brazilian Potash Institute, 1964.

SEGATTI, S. A **expansão da agroindùstria Sucroalcooleira e a questãoâo do desenvolvimento da microrregiao de Dracena-SP.** Dissertation (Masters in Geography), Presidente Prudente: FCT/UNESP, 2009.

SHIKIDA, P. F. A. Modernization of the sugarcane agroindustry during PROÀLCOOL: an exploratory approach using factor analysis. In: **Revista Economia & Empresa**, Sâo Paulo, v. 3, n.1, 1996, p.20-33.

Sugarcane Expansion in the Midwest: Limits and Potential. In: **Revista de Politica Agricola.** Year 1, n. 1. Brasilia, DF: Secretaria Nacional de Politica Agricola, Companhia Nacional de Abastecimento. April/May/June 2013 Year XXII - n. 2, p. 122-137.

SILVA, M. A. M. Bionergia e viabilidade da produção de alimentos para quem? In: **Revista Formaçâo**, n.15, v. 2, Presidente Prudente: Unesp, 2008 - p.198206.

SILVA, S. L. Q. da. The growth of sugarcane plantations in Brazil in the 70s. In: **Revista Brasileira de Economia**. Rio de Janeiro, 1983, p. 39-54.

SORJ, B. **Estado e Classes Sociais na Agricultura Brasileira**. 2ª Ed., Rio de Janeiro: Guanabara, 1986.

SOUZA, C. A. **Impacts on ethanol production**. 2014. Available at: http://www.scgms.com.br. Accessed on: Jul. 2013.

SOUZA, M. A. de. The territorial dynamics of sugarcane agribusiness and the agro-ecological zoning of sugarcane: notes for a debate. In: **revista campo-território**, v. 5, n. 10, 2010, p. 148-167.

SOUZA, S. M. R. de. **The emergence of the agribusiness discourse and the**

expansion of sugarcane activity: discursive strategies for the action of capital in the countryside. Thesis (PhD in Geography), Presidente Prudente: FCT/UNESP, 2011.

SZMRECSÀNYI, T. **O planejamento da agroindùstria canavieira do Brasil (1930-1975)**. Sâo Paulo: Hucitec, 1979.

et. al. **Dimension, risks and challenges of the current sugarcane expansion.** Brasilia: Embrapa, 2008 (texts for discussion 32).

TEIXEIRA, J. C; HESPANHOL, A. N. The Center-West region in the context of agricultural changes in the post-1960 period. In: **Revista eletrônica da Associaçâo dos Geógrafos Brasileiros - AGB**, Très Lagoas section, V.1, n° 3, year 3, May 2006, p. 52-65.

The trajectory of Brazilian cattle ranching. In: **Caderno Prudentino de Geografia**, Presidente Prudente, n.36, v.1, jan./jul. 2014, p. 26-38.

TEIXEIRA, J. C. **Changes in farming in Très Lagoas in the context of Mato Grosso do Sul.** Dissertation (Masters in Geography), Presidente Prudente: FCT/UNESP, 2001.

Modernization of Agriculture in Brazil: Economic, Social and Environmental Impacts. In: **Revista eletrônica da Associaçâo dos Geógrafos Brasileiros - AGB**, Três Lagoas section, Vol. 2, No. 2, year 2, 2005, p. 21-42.

TEIXEIRA, M. A. **Agricultural changes in Mato Grosso do Sul**: the example of Grande Dourados. Thesis (PhD in Geography), Sao Paulo: USP, 1989.

TEUBAL, M. The peasantry in the face of the expansion of agribusiness in Latin America. In: PAULINO, E. T.; FABRINI, J. E. (Org.). **Campesinato e territórios em disputa**. Sao Paulo: Expressao Popular, 2008. p.139-161.

THOMAZ JÛNIOR, A. **Por trás dos canaviais os (nós) da cana. The relationship between capital and labor and the workers' union movement in São Paulo's sugarcane agro-industry.** Sao Paulo: Annablume/FAPESP, 2002.

For a "*crusade*" against hunger and agribusiness - capitalism's new destructive agenda and the challenges of a time not postponed. In: **Revista Pegada**. Presidente Prudente: UNESP, vol. 9 n.1, 2008. Available at:< http://www4.fct.unesp.br/ceget/PEGADA91/02-9-1-ThomazJr.pdf> . Accessed on: June 2012.

TONIN, J. R; TONIN, J. M. From Proàlcool to "Próetanol": new challenges in Brazilian ethanol production. In: **Informe Gepec**, Toledo, v. 18, n. 1, jan./jun. 2014, p. 61-76.

VEIGA FILHO, A. de A; RAMOS, P. Proàlcool and evidence of concentration in sugarcane production and processing. In: **Revista Informaçôes Econômicas**, Sao Paulo, v.36, n.7, jul. 2006.

XAVIER, C. V. The expansion of the sugar-alcohol industry in the Andradina region and the crisis of capital reproduction. In: **Revista Eletrônica da Associaçâo dos Geógrafos Brasileiros** - Seçâo Três Lagoas/MS - n° 11 - Ano 7, Maio 2010.

; BOMBARDI, L. M. The expansion of sugarcane cultivation in the context of the development of agrofuels: transformations in land use and new labor relations. In: **Proceedings of the XIX NATIONAL MEETING OF AGRICULTURAL GEOGRAPHY**. Sao Paulo: USP, 2009, p. 1-16.

WEBSITES CONSULTED

www.agesul.ms.gov.br

www.agraer.ms.gov.br

www.agricultura.gov.br

www.ambiente.sp.gov.br

www.biosev.com

www.biosulms.com.br

www.bunge.com.br

www.copersucar.com.br

www.dsr.inpe.br/laf/canasat

www.embrapa.br

www.famasul.com.br

www.geoufsc.com.br

www.ibge.gov.br

www.ig.ufu.br/revista/caminhos.html

www.imasul.ms.gov.br

www.inpe.br

www.mda.gov.br

www.mst.org.br

www.nepo.unicamp.br

www.novacana.com

www.odebrechtagroindustrial.com

www.plataformabnds.org.br

www.raizen.com.br

www.scgms.com.br

www.semac.ms.gov.br

www.tononbioenergia.com.br

www.udop.com.br

www.unica.com.br

www.usinasaofernando.com.br

www.usinavicentina.com.br

www4.fct.unesp.br/ceget/PEGADA91/02-9-1

MIX
Papier aus verantwortungsvollen Quellen
Paper from responsible sources
FSC® C105338

Printed by Books on Demand GmbH, Norderstedt / Germany